IN DENIAL:
DARWIN'S
DELUSION

Max Gorman

ISBN-13: 978-1546669005
ISBN-10: 1546669000

CONTENTS

ACKNOWLEDEGMENTS

My special thanks to Anna Sullivan for her support and encouragement throughout the writing of this book.

My gratitude to Andrew Kean for reading the manuscript and assuring me that it was a worthwhile undertaking.

And thanks too to my son Ewan for urging me to get the book published at a time when I needed this stimulation.

I wish also to express my appreciation to Jill for her excellent typing and presentation of the manuscript and to Mike for all his varied work towards the production of the book.

Finally, I thank David Rosevear and the Christian Science Movement for kindly granting permission to include the illustrations of Diatoms from photographs by Frank Cousins.

INTRODUCTION

Now that all the hype from the media in connection with the centenary and a half of the publication of Charles Darwin's book *On the Origin of Species* is dying down, it is time to take a long hard look at what all the fuss was about and whether it was actually deserved.

That may seem a deliberately provocative statement. But it is serious. I, like an increasing number of others who have taken the trouble to think about it, regard Darwin's theory of evolution as a pernicious myth whose overblown significance has probably done more than anything else to undermine the culture of the West since it was first swallowed hook, line and sinker.

Though not by everybody. Many of Darwin's contemporaries were far from convinced by the ideas expressed in his book, and approached it with an intelligent critical scrutiny which has on the whole been absent until fairly recently. Nor – and this is significant – did most of this critique rest on a religious base, but from free-moving thinking, carefully and objectively evaluating the evidence that Darwin presented, and the validity of the premises upon which he operated.

Not that he presented much evidence. Much more than is commonly realised, plain belief, sometimes openly admitted, but mainly disguised (even perhaps from himself) by the sheer bulk of what purported to be 'observation' was the foundation of Darwin's thought, every bit as much as it is in the approach of those whose belief springs from religious conviction.

The only difference is that Darwin's was the 'god' of 'natural selection'. This is evident throughout *On the Origin of Species*. Basing his idea on the selected breeding of farm and domesticated animals, and his encounter with the remarkable variety of finches on the Galapagos Islands, he then felt constrained to apply it obsessively to everything he saw, flying cheerfully in the face of an almost total lack of evidence, stretching the notion to what can only politely be called absurd lengths. 'Natural Selection' becomes personified into an entity in itself rather than the accidental process it actually is.

We shall develop this point later. But what else for the moment are we to make of a remark like this:

> 'Daily and hourly natural selection is scrutinizing throughout the world, every variation, even the slightest, rejecting that which is bad, preserving and adding up all that is good.'

Origin of Species, pp.83-84

Or again:

> 'Man selects only for his own good. Nature only for the being which *she* tends...'

(My italics)

This is hardly the language to describe the impersonal action of chance environmental pressures upon equally chance variations of species. It would appear that, conscious of the inherent inadequacy of his theory to explain what he wanted it to, Darwin is forced to invest natural selection with a quality, a capacity, beyond what it is legitimately capable of bearing.

Nevertheless, while Darwin himself always regarded his idea as a theory, and therefore vulnerable to certain areas of disproof he was honest enough to identify, since his time the attitude of most biologists towards it has gradually hardened into what can only accurately be described as a dogma – akin to a religious dogma. This now means that for the vast majority of biologists it is no longer a theory but a fact. Nor is this the result of substantially more evidence than was available to Darwin. Let us compare the attitude of the producer of the theory with that of his more recent and contemporary advocates. Referring to the conspicuous lack of intermediates so necessary to his concept of gradual evolution from creature to creature in the fossil record, he says:

> '...the case at present must remain inexplicable, and may be truly argued as a valid argument against the views here entertained.'

Not only are they "views" but they are also "entertained". Implying that doubts can be too.

But not for Julian Huxley, as he makes clear at a conference in 1959:

> "The first point to make about Darwin's theory is that it is no longer a theory but a fact."

Or for Richard Dawkins:

> "The theory is as much in doubt as the earth goes round the sun."

If Darwin were with us today one wonders whether he would be entirely comfortable with Mr. Dawkins as a disciple. Or even the ostensibly more 'balanced' Mr. Huxley. The same goes for the majority of the Neo-Darwinian dogmatists. The theory cannot be any more scientifically proved according to the normal application of scientific method – as we shall show later – than the existence of God. Not that this matters – as we shall also explain later. But as Darwin's theory is always paraded *as* scientific, and its legitimacy asserted to rest on this foundation, we will be constrained to expose the fallacy of this claim.

Why Darwin's theory, which actually should not be dignified by the use of that term, for it is in reality merely a hypothesis unsupported by the kind of evidence and proof necessary for that designation, is held on to so tenaciously and uncritically by most of the scientific community that to question it is regarded as heresy is an interesting subject to examine, and we shall be doing just that.

Suffice it to say at this juncture that the main reason for this, and the petrification of the theory into 'fact' is the threat its overthrow constitutes not only to the self-image of scientists but of science itself. Science as a 'faith', as a sole and self-sufficient explainer of the Universe, is challenged and undermined by any idea that there may be areas of this Universe outside its boundaries. Thus, open-mindedness is literally out of the question.

Kuhn has given us a most valuable critique of the manner in which science develops in *The Structure of Scientific Revolutions* (1962). He shows how science progresses not continuously, as is usually assumed, but sporadically from paradigm to paradigm – the term he uses to describe the way an accepted mind-set functions to dictate a particular way of looking at things by the scientific community under its influence. The paradigm exerts a strong hold, a dominating influence upon its followers until eventually, after a long period of struggle, it is overthrown by a new paradigm – a process he calls 'paradigm-shift'. It takes a lot to overcome the resistance of the supporters of the reigning paradigm, who will understandably fight hard to defend what their research and employment depend on. They are identified with it and committed to it in more ways than one. Nor are the servants of the paradigm always conscious of their position. It might be more accurate to describe them as its victims!

Neither will the production of evidence in favour of the new paradigm, nor even the use of logic, necessarily prevail against ingrained and conditioned adherence to a paradigm imagined to be upheld for objective reasons.

Now Darwinian theory is a paradigm, and thus subject to all the xenophobia and resistance to challenge and change attendant on any paradigm. Worse still, it is not even recognised as a paradigm, so petrified has it become. But it is also connected with a larger, equally unrecognised paradigm. The paradigm of science itself.

This is why a concept like Intelligent Design constitutes such a serious threat to orthodox scientists. This is why they cannot afford to even entertain it. It challenges the very world-view upon which science rests. In a word, its materialism. It posits a cause for events outside its narrow confines. And a conscious cause at that. Science has never been very happy about consciousness. Let alone a consciousness of this order. A consciousness that can embrace pattern, purpose and design on such an infinitely intricate scale as permeates the wonderful web of nature.

So bearing in mind the self-inflicted myopia which incapacitates it as a mode of knowledge, except within the limits of its own parameters, it is all too understandable that science cannot but be blind to possibility of anything beyond the boundaries of its self-definition. Scientists have thus become prisoners of their own paradigm. Nor are they necessarily aware that they are in a prison. Its bars are not obvious to them because they are a set of implicit assumptions, conditioned and unexamined. There is no world outside, therefore they cannot be in prison.

Even when compelling evidence is produced which contradicts the validity of the paradigm, or indicates its parochialism, it makes no difference to its witting or unwitting supporters. This is partly because it cannot be assimilated to the paradigm. The conclusions it suggests are alien to it, to its very nature. But also because it challenges and undermines the very beliefs, and unbeliefs – which are also beliefs – 'religiously' held by its adherents. For atheism and scientism (and I use that word deliberately) are, as has been pointed out before, ideologies imbued with what is indistinguishable from religious fervour. Thus mind-lock ensues.

I shall be looking closely at a large body of evidence pointing to causal activity beyond the explanation of the Darwinian model, which projects inferences that are inescapable not only to the logic – which scientists pride themselves on but don't always use – but to ordinary common sense.

CHAPTER 1

DEFECTS IN DARWINIAN THINKING

Before we go any further let us look at Darwin's famous theory. It is only right to do so before criticising it in depth.

The first thing that strikes one is how curiously trite it is in view of the amount of applause it has received. We have looked at some of the reasons for this already in the Introduction – but there are others, and we will identify them shortly. But first the theory:

It could briefly be described as chance variation – nowadays called 'mutation' – operated upon by natural selection. This, Darwin thought, was enough to account for not only the differences *within* species but the difference *between* species. So he gave his book the title 'The Origin of Species'.

A hard look at the book – but the trouble is very few actually read it – reveals how false that title actually is. At no point in his book does Darwin ever show how a new species originates. A more truthful title would be 'The Preservation of Species' as the philosopher Ouspensky suggests. Or, as I would suggest, 'The Origin of *Varieties*'.

Every farmer knows that by selectively breeding animals for certain characteristics which he notices and wants, he can, after a number of generations, produce a better herd or flock. Successive selection for particular characteristic or characteristics and breeding only from such animals will ultimately result in a new breed or variety. Darwin, like most of us, was aware of this process, which is called 'artificial selection'. He admits that this was the germ of his theory of natural selection, the very basis of his thought. He affirms this clearly in his autobiography:

> "I came to the conclusion that selection was the principal cause of change from the study of domesticated productions; and then, reading Malthus, I saw at once to apply this principle." (67-8).

The comparison between artificial selection and natural selection is an analogy, and only an analogy. It is very important to remember this, for analogy was central to Darwin's thinking. He used it again to argue the relationship between the development of varieties and the development of something very different – species.

Now, analogy can, if properly used, be a valid method of thought. It can be very helpful. But as we all know, analogies cannot be pressed too far. There can be false analogies. There can be misleading analogies. However convincing they might superficially appear, they can be literally confidence tricks. They can induce

9

false confidence. A kind of sleight of hand.

Darwin was an honest man and was therefore probably not guilty of a deliberate confidence trick. Rather he so convinced himself of the truth of his analogies that he became a victim of his own confidence tricks. The persuader, and there is no doubt that Darwin *was* a persuader, persuaded himself.

Using his vast knowledge as a naturalist he evoked such an apparently convincing argument for his theory, albeit by widespread employment of special pleading, that the sheer weight of the information about plants and animals at his disposal makes the reader, unless he or she stays alert, imagine that he knows what he is talking about. And of course, he does – up to a point. That point is where he 'jumps the points' – by extrapolation from analogy.

Now, analogy as a mode of thought, even good analogy, is not science. Neither is it logic – a kind of thinking which science would like to claim it uses as a necessary aspect of the scientific method. Nor is it even 'common sense' – which scientists would like to believe they are vaguely associated with. Therefore, whatever kind of theory Mr. Charles Darwin fondly believed he had produced in his book *The Origin of Species*, it is not a scientific one. It is curious indeed that a theory contradicting so glaringly the requirements of legitimate scientific thought, has been accepted as scientific for so long by so many members of the scientific community.

What are the reasons for this? They have already been identified. Here was a theory which seemed to be plausible enough scientifically if you did not look too hard at it, and you did not do that precisely because it was just what you were looking for if you were an atheist who had always felt secretly uneasy that there might be more to the Universe after all than your narrow, one-dimensional conception of it could contain. This was the answer. Now those doubts could be done away with. No need to worry any more about the ubiquitous lineaments of design that stared you in the face. No need to wonder whether there was a Purpose behind nature and man, which one had sooner or later to come to terms with. Darwin had done it! Science was safe. Its paradigm had prevailed. Chance, garnished with a shake-up from the struggle for existence, was all you needed to do the trick. Better not look too closely at it. It might not be scientific after all!

If one does look closely at it, even cursorily in fact, one finds that over and over again Darwin does not prove – as one might reasonably expect a scientist to do – but *believe*. Though his book, with its wealth of detail and its apparently carefully developed theme, presents itself as an extended argument, even a fairly conscious critique of it will reveal it to be a masquerade for a series of beliefs, or non-beliefs which are of course *also* beliefs.

I would argue that analogy as used by Darwin is in itself a form of belief – though disguised as a form of legitimate thinking. One is asked to believe that something is like something else – when one can see they are different. 'As this – so this' appears to be a logical sequence or progression. In reality it isn't. Unlikeness is thus disguised as likeness.

Darwin resorts to analogy at two central, pivotal, points in his thought. The first is his comparison between artificial and natural selection. The second is his extension of the concept of the role of natural selection from variation *within* a species to variation *between* species, or as he liked to put it, the "origin of species." Both concepts are inherently flawed – as I shall show. But the greater defect by far lies in Darwin's attempt to blur the obvious distinction between a variety and a species. One might be able to accept that selection, natural or otherwise, could over a period of time result in the arrival of a number of varieties of pigeon. But that a frog could one day become a fox requires a leap of the imagination somewhat greater than a frog could perform!

Darwin was admittedly on fairly safe ground when he talked about pigeons – as he does at considerable length in the first chapter of his book *Variation under Domestication*. He was more than a bit of a pigeon fancier himself, and significantly his book was preceded by three years' rabid immersion in the society of pigeon breeders.

> "I have kept every breed. I could purchase or obtain...I have associated with several eminent fanciers, and have been permitted to join two of the London Pigeon Clubs."

– Darwin proudly states. Perhaps he would have been well advised to restrict his book to this field.

Returning to the role of belief in Darwin's exposition, it is in fact to his credit – as a human being, of course, rather than as a scientist – more openly displayed in other aspects of his attempt to establish his theory than its less obvious presence under cover of analogy. His 'argument' is in fact imbued with, and founded on, belief and supposition at almost every turn – certainly every major turn. I am not alone in this observation. Listen to Adam Sedgwick, the learned geologist and contemporary of Darwin, commenting in his article 'Objections to Mr Darwin's Theory of the Origin of Species" (*Spectator*, 1860):

> "We maintain that neither Mr. Darwin nor any other writer has yet given the slightest proof that the recognised discontinuities in species can be accounted for by the action of secondary causes. Indeed our author makes at any time but little use of the word 'to prove' in any of its inflections. His formula is 'I am convinced', 'I believe' and not 'I have proved'. We are not finding fault with these more modest forms of expression; but we may be allowed perhaps to remark, that they are the formulae of a creed, and not of a scientific theory."

And Mivart makes the same observation:

> "We feel we now have a right to demand that they (his conclusions) shall be proved before we assent to them...This is the more necessary, as the Author, starting at first with an avowed hypothesis, constantly asserts it as an undoubted fact, and claims for it, somewhat in the spirit of a theologian, that it should be received as an article of faith. Thus the formidable objection to Mr. Darwin's theory that breaks in the chain cannot be bridged over by any extinct or living species, is answered simply by an appeal 'to a *belief* in the general principle of evolution'." (Mivart's italics)

Jackson Mivart was a prominent biologist who wrote '*The Genesis of Species*' in 1871. He accepted 'evolution' but not natural selection.

The particular belief of Darwin that is referred to here by both men is his belief in 'the intermediaries' – the great host of plants and animals who are supposed to provide the connecting links between the major classes of species and thus reveal the continuity in the chain of evolution which Darwin required for his theory of gradual change of species into species to be true.

Unfortunately for Darwin, none can be found – either alive in nature today, or in the fossil record. The evidence of the record is in fact to the contrary. Over and over again it is marked not by any gradualism but by the sudden and inexplicable arrival of distinct new species.

The gaps in the record, or more properly, the record *of* the gaps, is a serious matter, a very serious matter indeed for the validity of Darwin's theory. Without signs of the necessary intermediates, the whole edifice would be a colossus with feet of clay. Clay and only clay minus the fossils – that was the trouble. As he admitted in a letter to Asa Gray in 1857:

> "...one's imagination must fill up very wide blanks."

And in his book:

> "...the case at present remains inexplicable and may truly be argued as a valid argument against the views here entertained."

That is an honest confession of ignorance. They are 'views', and they are 'entertained'. He is well aware that it is not a cast-iron theory. It would be well if even a shadow of the doubts of its producer assailed some of its present-day advocates.

As he says in *The Origin of Species* (Chapter 9): On the assumption that his theory is correct:

> "...the number of intermediate species which have formerly existed on the earth must be truly enormous. Why then is not every geological formation and every stratum full of such intermediate links? Geology assuredly does not reveal any such finely graduated organic chain; and this perhaps is the most obvious and gravest objection that can be urged against my theory."

Darwin's answer is to resort to belief:

> "The explanation lies, as I *believe*, in the extreme imperfection of the geological record." (My italics)

And that too is the explanation offered by Darwinians in general ever since – despite the fact that not only was the fossil record very much better in his day than Darwin liked to think, but that now, 150 years later, despite all the increasingly expert search that has taken place from that day to this, the picture has not been changed. The hoped-for intermediates have not been found. Whether then we are dealing with belief, or wishful thinking in this matter, is something to be left to the reader. Regrettably, it isn't science.

The leading palaeontologists of Darwin's day were well aware that the fossil record was not perfect, but this did not affect their confidence in the adequacy of the record as evidence for its major outlines. One of these, John Phillips, gave the Reid lecture at Cambridge in 1860 on 'Life on the Earth: Its Origin and Succession' in which he reviewed the current evidence of palaeontology in view of Darwin's theory. He maintained that Darwin had grossly overstated the case for the imperfection of the fossil record. Whatever its defects, nevertheless in broad outline, and particularly for shell-bearing marine animals, it was good enough to test the plausibility of Darwin's belief in extremely slow trans-specific changes. Not only was there no positive fossil evidence for these transitions but also, more seriously, it was now clear that the earliest known forms of Palaeozoic life were already highly complex organisms.

If life had evolved into its wonderful profusion of creatures little by little, one would expect to find fossils of transitional creatures which were a bit like what went before them and a bit like what came after. But no one has yet found any evidence of such transitional creatures. This oddity has, as we have said, been attributed to gaps in the fossil record which gradualists expected to fill when rock strata of the appropriate age had been found. However, in the last few decades geologists have found rock layers of all divisions of the last 500 million years, and no transitional forms were contained in them.

One final point, that has been touched on earlier – but needs expanding. Though on the one hand Darwin purports to present natural selection as a purely mechanical thing, where chance variation – or chance mutation as they call it today – is acted upon by equally chance pressures from the environment, the chance, resistance to which by those creatures with the chancely-provided advantageous variation enables them to be selected for survival, he seems to have understood, at some subconscious level, that the whole thing wouldn't work – or at best wouldn't get you very far. That there was little mileage in it beyond intra-species variation at any rate. So he has to invoke a 'personality'. A 'Selector' slyly appears; by a sleight of hand that may have deceived Darwin himself. Instead of the automatic process which he ostensibly proposes, unguided by God or other designer, which is supposed to be his main argument, he comes out with statements like this:

> 'Daily and hourly natural selection is *scrutinizing* throughout the world, every variation, even the slightest, rejecting that which is bad, preserving and adding up all that is good.' (*Origin of Species*, 83-84) (My italics.)

A 'Scrutinizer' has appeared. Is this Nature or natural selection? In either case, chance – which is all 'the survival of the fittest' depends on – has quietly disappeared, and so has the purely scientific basis of the theory. Again, listen to Darwin's tone as he compares artificial with natural selection:

> 'Man selects only for his own good. Nature only for the being which *she tends*...Can we wonder then that nature's productions should be far "truer" in character than man's productions, that they should be infinitely better adapted to the most complex conditions of life, and should plainly bear the stamp of far higher *workmanship*?' (My italics)

This is not the language to describe the hit-or-miss action of mindless mechanicality that fortuitously forges the changes in animals or plants that the pure and hard process of unaided natural selection is equipped to result in. It is almost reverential in tone. Nature has become a 'Person' and is substituted for natural selection which then becomes 'her' agent under her conscious control:

> '...she may be said to scrutinize with a severe eye...every habit, instinct, shade of constitution. There will here be no caprice, no favouring: the good will be preserved and the bad rigidly destroyed.'

Nature is thus presented as acting steadily, justly, and with virtually divine discernment, separating the good from the bad. It is pseudo-Biblical expression. Nature looks strikingly like God's surrogate. Darwin is no longer, if he ever was, a scientist. One suspects that he never really convinced himself that mere natural selection could do what he said it could do – unaided.

But does it matter? Yes, if Darwin's theory is to be accepted, as it is and has been, as scientific. No, if it is recognised as it is – a belief.

CHAPTER 2

THE INDELIBLE

DISTINCTNESS OF BEINGS

What prompted me to write this book was what I can see without looking further than my little garden. One does not have to wander the world to learn what there is to learn – a walk around the garden will do.

Even from where I am sitting I can see various and wonderful beings. Some are plants, and some are animals. All endowed with the mystery of life, yet all so distinctively different. Bird and bumble bee, apple tree and acacia, beetle and butterfly, fern and fennel, lily and lichen, fly and ant. They're all there. And utterly themselves!

Each kind of creature is so distinctively moulded, so specially shaped, so supremely unique. It stares one in the face. A face that is equally distinct.

Here are no loose ends, no malformations, no stragglers. No imperfections litter the landscape. There are no blurrings of the species as Darwinism would lead us to expect.

On the contrary, species are special, as the word implies, everywhere triumphantly proclaiming their intrinsic and iridescent individuality! The bumblebee bumbleth as only a bumblebee can, imbued with its bumblebee-hood, its essential nature. The sparrow chattereth in its own special language. The dragonfly dallyeth over the little pond, declaring, 'I am the king of flies, listen to my royal rustle.' The humble worm tunneleth, saying, 'I am no snake – but there is none like me, doing what I do.' And there, upon the wall the ivy creepeth, crying, 'I am I and not another thing.'

They're all different, and their selfhood shines through. This is something we all see. It is, there before our eyes. Here is no continuity, no blurring of distinctions between creatures – but self-containment, indelible individuality, and utter variousness.

And, we, mankind, have always seen this. It calls for no great perception. Aristotle saw it. Plato saw it. The author of Genesis saw it. That great observer of nature, Linnaeus, thought it too obvious for argument. All naturalists except the Darwinians have seen it. I've seen it – you've seen it.

Actually I'm convinced Darwinians themselves see it. After all, they are supposed to be observers of nature, and though their capacity in this respect is somewhat lacking, or lacklustre in comparison with most of the naturalists of the past and present – due probably to the dulling effect of their doctrines upon their powers of visual perception – they are not blind. It is more likely they are deliberately perverse. They refuse to face the facts patently perceptible before them. They won't believe what their eyes can see. It's not a matter of 'I'll believe it when I see it', but rather

'I'll *see* it when I believe it.'

It is the Darwinians rather than the anti-evolutionary biologists that have imposed *a priori* principles upon their perception of nature. It is quite untrue to say that those who opposed the Darwinian thesis at the time did so because of the religious beliefs they may or may not have had. When we look at how they actually approached the matter it is absolutely clear that their conceptions came not from belief – as their opponents liked, and like, to think – but from direct observation and common sense. Their approach was empirical not theological.

Let us look at the views of some of the leading representatives of what has been called the 'typological' model of nature. Typological, because they noticed that all classes of plants and animals inhabiting the planet were essentially different and distinct. There was no gradual continuity to be found between them, no intermediates as required and wished for in vain by advocates of the Darwinian model. On the contrary, nature exhibited a number of different *types*, clearly identifiable by their possession of a cluster of special characteristics unique to a particular class of being. So when the species of one class, say birds, were compared with any non-avian species, all could be seen to be equidistant in terms of their fundamental avian characteristics – so that no species of bird is fundamentally closer to any non-avian species.

No one has any difficulty in recognising a bird, whether it is an eagle, an ostrich, or penguin. Or a cat, whether it is a domestic cat, a lynx, or a tiger. The species may be different, but the class is self-evidently the same. The characteristics of 'cat-hood' can be seen at a glance. Furthermore, no one can name a cat or a bird that is not fully characteristic of its class. No bird is any less a bird than any other bird. Nor is any cat any less a cat or any closer to a non-cat species than any other cat.

Birds possess a number of characteristics that are absolutely unique. Their wings are formed of those wonderful things – feathers. A unique and sophisticated arrangement of flight feathers on the wings form the basis of what is clearly an aerofoil of variable geometry so that it has the ability to vary the shape and aerodynamic properties of its wing at take-off, landing, and for various sorts of flight – flapping, gliding soaring. I could go on. Birds also possess a unique continuous lung system which enables them to breathe in a particular way necessary for the exertion of flight. This maintains a continuous air-flow – quite different to the breathing in and breathing out of other animals and man.

A similar suit or set of unique defining characteristics can be assembled for a host of other classes of animals and plants – from insects to flowering plants. It is in fact the *norm* throughout the whole of nature.

As far as individual defining characteristics are concerned one could continue citing a great many which are without similarity or precedent in any other part of the living kingdom, and most importantly are not led up to in any way through a series of transitional structures. Such a list would include the spinneret of the spider, the jumping organ of the click beetle, the wing of a bat, and countless more. In these and other ways these creatures stand in splendid isolation from all other beings of the natural world.

For Richard Owen, first Superintendent of the British Natural History Museum, and a 'typologist', there was simply no evidence that the sort of gradual evolution by natural selection postulated by Darwin had ever occurred. As he wrote in his article for the Edinburgh Review 'Darwin on the Origin of Species':

> 'Is there any *one* instance proved by observed facts of such a transmutation? We have searched the volume in vain for such. When we see the intervals that divide most species from their nearest congeners, in the recent and especially the fossil series, we either doubt the fact of progressive conversion or as Mr. Darwin remarks in his letter to Dr. Asa Gray, "one's imagination must fill up very wide blanks." The last ichthyosaurus, by which the genus disappears into the chalk, is hardly distinguishable specifically from the first ichthyosaurus, which abruptly introduces that strange form of sea-lizard in the Lias. The oldest Pterodactyl is as thorough and complete a one as the latest. No contrast can be more remarkable, nor, we believe, more instructive than the abundance of evidence of the various species of ichthyosaurus throughout the marine strata of the oolithic and cretaceous periods, and the utter blank in reference to any form calculated to enlighten us as to whence the ichthyosaurus came, or what it graduated into, before or after those periods.'

To this day the 'utter blank' remains. As does the utter blank on the faces of the Darwinian fundamentalists when they are forced to acknowledge such uncomfortable facts. For whether they like it or not, fundamentalists they are.

Like many another creature, the Pterodactyl and the Ichthyosaurus comes... and then goes. It appears – doubtless plays its role in the scheme of life on this planet – and then disappears, without trace. With superb indifference to any putative continuum so necessary to Darwinian theory. As indeed do many other beings.

Richard Owen, then, was clearly not criticising Darwin's theory from a religious position. He was simply looking at the facts as they stood – and still stand – despite another century and a half of geological and paleontological investigation.

Nor was H.G. Bronn, the leading German palaeontologist of the nineteenth century. In his 'Review of the Origin of Species' he found "the enormous gaps which now confront us in the series of plant and animal forms" impeded his "complete consent".

Francois Jules Pictet too, another leading continental palaeontologist, also found the crucial empirical evidence of intermediates absent. In his response to Darwin's book he writes:

> "Why don't we find these gradations in the fossil record, and why, instead of collecting thousands of identical individuals, do we not find more intermediary forms? To this Mr. Darwin replies that we have only a few incomplete pages in the great book of nature and the transitions have been made in the pages we lack. But why then and by what peculiar rules of probability does it happen that the species which we find most frequently and most

abundantly in all the newly discovered beds are in the immense majority of the cases species which we already have in our collections?"

Richard Owen, Jules Pictet, H.G. Bronn and a number of others were speaking as experts in their field, and greater experts than Darwin in making these observations. Their statements do not come from any religious standpoint any of them may have possessed. They did not leap to the creationist model. They were simply forced to these criticisms by looking impartially at the sheer weight of the factual evidence for them – and the striking lack of any counter-evidence.

What then is the significance of the non-existence of intermediates over such colossal periods of time, their absence in the geological record? Simply this. It endorses the concept of the intrinsic distinctity of creatures, their unwavering self-sameness, the strength of their identity – preserved through vast periods and varying environmental pressures. Their difference is a deep and enduring thing.

Let us now look closely at some examples of what could be called 'the indelible distinctness of beings'.

We'll begin with a miniscule creature as humble as it is entitled to be proud because of its role in nature, including your very ability to breathe, dear reader, as you read this page. Although you may not be aware of its existence, let alone any connection it has with you.

We'll begin with it because it was here at the beginning, and has been with us ever since. It's a very tiny plant called the blue-green alga. It looks small, having only one cell, but what it does is big. This little creature is the first photosynthesiser. The first to produce oxygen, upon which all animal and human life on the planet depends. What the marine blue-green alga did was to gradually change a carbon dioxide-rich atmosphere into an oxygen-rich one. Thus, laying the foundation for all subsequent life on Earth, both aquatic and later terrestrial. This extraordinary being – for photosynthesis *is* extraordinary – its arrival is inexplicable, and being is more extraordinary – led the way, helping to create a habitat suitable for that sparrow on the branch, that cat on the wall, that wasp on the window-sill, and a place for you and me. Even today the blue-green alga produce about 60 per cent of the oxygen in our atmosphere. Not too bad.

But that's not what I want particularly to talk about at the moment. What I want is to point out the unwavering unchangingness of this minute and delicate being over vast and unimaginable stretches of time. I mean, about 3 billion years – the date of its earliest fossils. Now the interesting thing is that comparing the shape of these ancient imprints with the contemporary blue-green alga, there is hardly any difference between the two. It's scarcely changed a bit through all the vicissitudes and environmental variations of three thousand million years. That's a long time. That's a long time, a very long time, to retain your identity. It indicates an innate self-sustaining potency – centrifugal rather than centripetal in nature.

Now look at the dragonfly. It too hasn't changed a bit since it made its debut 380 million years ago. Its fossil dated then shows that it was essentially the same when it hovered over that primeval swamp, as the one you can see hovering over that pond today!

The same applies to many other creatures, both animals and plants. That fern I can see over there in a shady spot of the garden has been with us and before us a long, long time in all its fair, fine frondiness. It and its family dominated those dark, dank carboniferous forests for millions of years – and there it still is today! Those ferny forests with their multitude of tree-ferns, and many other varieties, overcast the planet from 345 million years ago to about 285 million years ago. Haunted by many a strange creature we don't know anymore. But look – the fern has come through!

Starfishes showed up suddenly in the Cambrian period about 500 million years ago without any advance warning whatsoever, and have been with us ever since. So did jellyfishes and sea urchins, together with a number of sea snails – and they all seem to like themselves as they are!

The bat, with its special kind of wings and its unique and utterly extraordinary echolocation faculty, is another creature that appears – without any predecessor – sixty million years ago, according to the fossil record, and insists on remaining the same old bat we can see in the belfry to this very day. It hasn't changed a bit.

Ants too, preserved in amber from the carboniferous forests 320 million years ago, are very much the same ants you see going about their business in your garden today. And in the Konigsberg fossil collection of over eleven thousand ants in amber you can see several milking aphids like 'cows' as they all still do.

I've been holding back on the coelacanth – which itself has been holding out on us for about 72 million years, when it was thought to have become extinct by all the best Darwinians. But it unnerved and disappointed them when one was hauled up in 1938 by fishermen in the Indian Ocean, off Cape Province in South Africa. Unnerving because it had no business to be there at all! No business to cock a snoop at the fossil record and the much-respected authority of the geological column. Disappointed because, on the basis of the many fossil coelacanths found and described up till then, it was firmly believed that this fish was a hoped-for missing link between fish and tetrapods – in particular, the amphibians. This belief had become 'evolutionary fact' – still doggedly perpetuated in textbooks.

But the coelacanth let them all down on this too. It declared itself a fish and nothing but a fish. Its fins were not incipient legs, as everyone had hoped – or decided. They were fins, pure and simple. Like any other fish, the coelacanth swam – it didn't crawl on the ocean floor as it was supposed to. What an insult!

However, the real point about the coelacanth is not that it refused to be a missing link – but that it firmly refused to be anything but the fish it was, from its arrival all those millions of years ago to what it was when it was sadly scooped up by those fishermen in 1938 – and doubtless remains today!

The same applies to all the creatures just mentioned, and many more that could be. In fact this essential unchangingness, apart from minor variations over millions of years, is true of every plant and animal we know.

The significance of this, the longstanding constancy of creatures, is that it lends support to the belief held by a number of thinking opponents of Darwin, that there is an inherent 'inbuilt' limit to the amount of variation a creature can undergo if left to itself or not left to itself. Not merely a resistance to change, but an affirmation of its especial selfhood – with which it is innately imbued. One could call it a centre of gravity, even a 'specific gravity' against whose force a possible variation of the species cannot pull beyond a certain 'specified' (prolonging the concept) boundary.

One of the ablest and most acute critics of Darwin's ideas, Henry Fleeming Jenkin, expressed this observation in a slightly different way, and based it quite legitimately on the very evidence – the selective artificial breeding of domestic animals – which Darwin used in his attempt to justify his own theory. Let us have a look at it.

In a much-neglected article in The North British Review, June 1867, Fleeming Jenkin produced this interesting concept:

> 'A given animal or plant appears to be contained, as it were, within a *sphere of variation*; one individual lies near one portion of the surface, another individual, of the same species, near another part of the surface; the average animal at the centre. Any individual may produce descendants varying in any direction, but is more likely to produce descendants varying towards the centre of the sphere, and the variations in that direction will be greater in amount than the variations towards the surface. Thus, a set of racers of equal merit indiscriminately breeding will produce more colts and foals of inferior than of

superior speed, and the falling off of the slower will be greater than the improvement of the select. ...The tendency to *revert* is generalized in the simile of the sphere here suggested.' (My italics.)

He continues:

'If in every case we find that deviation from an average individual can be rapidly effected at first, and that the rate of deviation steadily diminishes till it reaches an almost imperceptible amount, then we are entitled to assume a limit to the possible deviation.'

Jenkin's conclusions, drawn from a careful study of domestic selective breeding of varieties of plants and animals, support the concept of the innate 'strength of identity' of beings which I have endeavoured to develop above. To return to him:

"Darwin's theory requires that there shall be no limit to the possible differences between descendants and their progenitors, or, at least, that if there be limits, they shall be so great a distance as to comprehend the utmost differences between any known forms of life. The variability required, if not infinite, is indefinite. Experience with domestic animals and cultivated plants shows that great variability exists. Darwin calls special attention to the various fancy pigeons, which, he says, are descended from one stock; between various breeds of cattle and horses, and some other domestic animals. He states that these differences are greater than those which induce some naturalists to class many specimens as distinct species. These differences are extremely small as compared with the range required by his theory, but he assumes that by accumulation of successive differences any degree of variation may be produced. He says little in proof of the possibility of such an accumulation, seeming rather to take it for granted that if Sir John Sebright could with pigeons produce in six years a certain head and beak of say half the bulk possessed by the original stock, then in twelve years this bulk could be reduced to a quarter, in twenty-four to an eighth, and so farther. Darwin probably never believed or intended to teach such an extravagant proposition, yet by substituting a myriad of years for that poor period of six years, we obtain a proposition fundamental to his theory. That theory rests on the assumption that natural selection can do slowly what man's selection does quickly; it is by showing how much man can do that Darwin hopes to prove how much can be done without him. But if man's selection cannot double, treble, quadruple, centuple any special divergence from a parent stock, why should we imagine natural selection should have that power?

When we have granted that the 'struggle for life' might produce the pouter or the fantail, or any divergence man can produce – we need not feel one whit the more disposed to grant that it can produce divergences beyond man's power. The difference between six years and six myriads, blinding by a confused sense of immensity, leads men to say hastily that if six or sixty years can make a pouter out of a common pigeon, six myriads may change a pigeon into something like a thrush."

Professor Fleeming Jenkin makes some telling points in his acutely argued critique of Darwin's analogy between domestic and natural selection. And he is right in stressing that this is a proposition fundamental to the whole theory. No matter how different the thirteen Galapagos finches might be – they were all finches.

Interesting too is his psychological observation that the effect of attempting vast geological periods of time may have, for some, a tendency to blur their clarity of thought in these matters and confuse their better judgement. It is another instance of the sleight of hand that Darwin uses to bamboozle his readers into accepting his theory.

But perhaps I am being a little unkind. In a sense, Darwin was not deliberately dishonest. He had at first to bamboozle himself.

CHAPTER 3

INEXPLICABLE COMPLEXITY

I'd *like* to start at the beginning of what is called The Universe – but, regretfully, cannot. The so-called 'Big Bang' theory, which is the name Fred Hoyle gave to what Georges Lemaitre, the Belgian priest who first produced it, had called 'the hypothesis of the primeval atom' or 'the explosion of the Cosmic Egg' about 13 billion years ago, obviously doesn't describe the beginning of The Universe. Something was obviously there *before* that in the first place, to have caused or produced this event. Something was going on behind the scenes to have created the 'cosmic egg' and timed its irruption at that particular occasion. Something does not come from nothing. Even rabbits come from hats! It is noteworthy that his theory in no way affected Abbé Lemaitre's belief in God. Nor Einstein's who, initially disagreeing, came to accept it. Fred Hoyle of course never did, holding on to his 'steady-state' theory to the end of his life on Earth. But he did come to believe in a Creator in his later years simply by using that faculty much underemployed by scientists –

common sense. As he said:

> "A common sense interpretation of the facts suggests a super intellect...and that there are no blind forces worth speaking about in nature."

And also:

> "The notion that the operating program of a living cell could be arrived at by chance in a primordial organic soup is evidently nonsense..."

It is generally accepted that such ideas prevented Hoyle from being awarded the Nobel Prize, while his colleague and collaborator William Fowler did receive this honour.

But back to the 'beginning'. Somehow or other a 'point of intense concentration of energy and matter explodes', and somehow or other stars and galaxies fly outward forming the ever-expanding entity we like to call the Universe.

Now, the term 'explosion' cannot describe this. It implies something chaotic and chancey. Whereas what we can actually see that has emerged from this process, which looks more like a controlled, directed effusion, has all the characteristics of order and pattern. Wherever we look, however far or near, we see the great pattern of galaxy, star, solar system, planet and moon. Within each galaxy we witness the intricate gyration of stars round galaxy, planets round stars, moons round planets. Each has its place, ever-turning within the vast tapestry. The overall impression is one of orchestration and order. And order implies an Orderer. While pattern implies a pattern-maker. And probably a purpose.

The spacious firmament on high,

With all the blue ethereal sky,

And spangled heavens, a shining frame,

Their great Original proclaim.

Addison was right. Order does not come out of chaos. Not on its own, anyway. And under the shining frame must lie a hidden frame – upholding it from within. 'Ex invisibiliis visibilia' – the visible springs from the invisible – as Eringena and many other ancient philosophers were convinced.

The First Life on Earth

The earliest living things on Earth are the bacteria. There are traces of them in the fossil record going back to over three and a half billion years ago.

Bacteria are, as we all know, very small, single-celled creatures and because of this they have been denigrated by scientists as 'primitive' etc. so that they can be placed as low as possible on the evolutionary scale to support the idea that life came initially from inorganic matter in some way or other as a result of something or other taking place in some kind (or other) of 'soup' a long time ago. The idea is to bring the inorganic and the organic as close together as possible in order to give the appearance of a continuum between the two. At one time viruses were supposed to do this, being simpler than bacteria. But unfortunately they refused to play the game by turning out to be all parasites, unable to exist without an infinitely higher developed host. So they must have arrived very much later in the evolutionary day. They're troublemakers. They play havoc with Darwin's theory.

Back to bacteria then! They've been called 'simple', but in reality there's nothing simple about them. They're as mysterious as anything else alive – anything else endowed with that mysterious quality called life. Though a bacteria cell might be single, it's one complex palpitating whole, eating, absorbing, breathing, reproducing, moving, and most importantly – sensing. It has that thing that materialists (many of whom are politely called Darwinists) can't explain, consciousness. Not like yours and mine. of course, not like that of a fish, or a bird, or a bear – but all the same it's there, that extraordinary knowingness, that awareness of what it's up to which comes from goodness knows where.

And if life means anything at all, no matter how wonderful its physical manifestations, it above all means that miraculous thing, consciousness. It should be the evolution of consciousness that should interest us – something entirely omitted by Darwin and his theory, but which should surely be the primary focus of our approach to life rather than only its more obvious expressions.

Bacteria can sense and respond to all kinds of things. Many sense a change in the concentration of substances in the environment around them including oxygen, carbon dioxide and various sugars. They also sense a change in acidity. They can move towards or retreat from areas of the environment depending on whether its contents or nature is beneficial or not to them.

While swimming, they keep comparing a stimulus with its previous value to determine whether it is increasing or decreasing – and thus whether or not to continue in the same direction. This indicates that they can actually remember a past state. In other words, bacteria have a memory.

Some bacteria can even sense the direction of a magnetic field, including that of the Earth itself. In water they use this ability to swim along inclined magnetic lines of force to reach the environment that favours them. Not sure I can actually do that myself!

I forgot to mention the extraordinary flagellum – the microscopically tiny hair-like 'tail' which enables many, though not all, bacteria to move through the surrounding medium. It's a very complex thing. It's no good saying it came about by chance, etc. The more this feature is studied, the more sophisticated it appears to be. Among other things it's the only structure in the entire planetary living kingdom which possesses and exhibits a true rotary motion. Unlike cilia which beat by the propagation of a wave from their base to their tip, the helical filaments which comprise the bacterial flagellum rotate rapidly like propellers and are driven by a reversible 'motor' at their base.

Furthermore, and central to our argument, the bacterial flagellum and the rotary motor which drives it, are not led up to gradually by a series of intermediate structures – and, as is usually the case, it is impossible to envisage a hypothetical evolutionary sequence of simpler rotors through which it might have evolved gradually. There's a lot for the Darwinist to explain and it's understandable that they very sensibly steer clear of this and many similar subjects.

Another significant thing about bacteria is that it has fairly recently been discovered that there are two distinct kinds: the Archaebacteria and the Eubacteria. In fact they are about as different between themselves in various characteristics as bacteria in general are from all other creatures. Hence the different names they have been given.

Thus, even at this stage of life we witness a major difference of species that cannot be accounted for by natural selection. Small changes in one direction possibly. But a fundamental gap like this is not something to be explained away so easily and simplistically.

Bacteria are also extremely adaptable. That word is used deliberately. It does not mean 'adapted' by chance or chance processes. I mean there is something else going on.

Today you'll find them all over the place, adapted to almost every possible – and virtually impossible – ecological habitat. As every gardener knows, or should know, they live in the soil, particularly the topsoil, about 30 billion to the ounce, or for the farmer half a ton to the acre, where they perform the indispensable role of mineralising dead matter. They're abundant at the bottom of oceans, and lakes – doing the same thing for the aquatic ecosystems. They can live at near freezing temperatures, and also in environments as hot as 110 degrees Centigrade. Some can tolerate high salinity – they're the only form of life to be found in the Dead Sea. Some bacteria have been found doing well at the base of the stratosphere, five miles above the earth. In other words, they get around – or are fairly ubiquitous, if you prefer: Of course we are talking about many different species – each adapted specially to its particular environment. Though interestingly they are found only in three distinct shapes: spherical, rod-like, or curved. Why, one wonders, has chance restricted itself to only these definite three?

And how can the lottery of chance variation and the survival of the fittest account for the extraordinary fact that time after time, everywhere they went, no matter what the conditions of the environment, no matter how adverse, these little creatures were always able to adapt? And why *should* they adapt? Weren't they doing very well as they were – in their normal, original environment? Why leave it? Why make things difficult for themselves? What advantage, survival or otherwise, could there be in entering invading areas, adverse or hostile to them? Why make the uninhabitable habitable, when you are doing very nicely, thank you, where you are? And how are we to believe, as the Darwinists would like us to, that over and over again chance would produce the whole succession of mutations necessary for so many bacteria to be a match for so many differently difficult areas of the planet – to reach the stage of being perfectly equipped to survive in what would initially be the unfittest of environments for their survival? Is not perhaps a helping hand necessary to explain this phenomenon?

Photosynthesis

We come now to consider that extraordinary process which plants perform which we have all heard of but rarely pause to appreciate as one of the marvels of life on this planet – photosynthesis – which literally means 'putting together with light' from the Greek '*photos*', light, and synthesis putting together. This is the activity of the blue-green alga without which, as has been said, all subsequent planetary life would not be here.

Photosynthesis is a process of unbelievable complexity, as continued scientific research has revealed, and much of it is still not 'understood'. A large number of different steps and stages are involved, all co-ordinated to the end of providing a plant with the energy, ultimately from the sun, to live and fashion the fabric of its body. Much, much more takes place in photosynthesis than the intake of carbon dioxide from the atmosphere and the release of oxygen into it.

Books have been written about it. Professorships have been devoted to it. The more it is investigated, the more the enormity of it is revealed and the more it is realised how much is unexplained within the actual process. But of course the process as a whole, as an entity, that it should exist at all, is not explained any more than the existence of the plant that depends on it. Photosynthesis is a phenomenon which tends to be taken for granted. It assuredly should not be. We should stare at it in amazement. For without it we wouldn't be here – to stare.

Which brings me to my second point. And one which is even more overlooked than its extraordinariness. It is the ecological one – and the unmet challenge it presents for the theory of 'natural selection'. Let me explain.

As any ecologist will tell you – if you haven't worked it out for yourself anyway – all animal and human life on this planet depends on plants for their ultimate food – and survival. Plants are called 'producers' and all other creatures 'consumers', by ecologists. You can see why. Plants can produce their own food – directly from the atmosphere through photosynthesis and from minerals in the soil. We, and the animals, can't. We need plants to do it for us. They're the primary producers of the food chain. They're its base. We, and the animals, either eat plants directly, or indirectly by eating animals. So we're consumers. All right, you knew all this. But hang on.

Here's my point.

It is generally accepted that animals, beginning with the zooplankton of the ancient ocean, 'evolved' *after* the plant plankton, in particular the blue-green alga, for the reason just stated – that they relied upon them for food formed through photosynthesis. They also relied on them for oxygen – without which animals can't exist and therefore could not come into being. The early Precambrian atmosphere consisted almost entirely of nitrogen and carbon dioxide. So oxygen had to be built up to a certain level before animals could arrive.

The question is: why, in terms of the theory of natural selection, should there be any further 'evolution' than that? Why should animals – creatures that were in the totally vulnerable position of not being able to photosynthesise, and utterly dependent on those beings, namely plants, who *could* do this, come into existence in the first place – being totally unfit in this respect for *any* kind of survival, let alone the survival of the fittest? There was clearly a 'stoppage' here. There is no continuum, in terms of natural selection, between plants and animals. Everything should have come to a stop with the early plants.

Put it another way. Why should life evolve any further? They were completely self-sufficient and managing perfectly well as they were. The loss of photosynthesis means the loss of self-sufficiency. It means a dependency-situation which is the very antithesis, the very reverse of the survivability which is the central feature of the theory of natural selection. It runs counter to its object. For obviously it is the self-sufficiency-situation which is the best for the survival of the species etc. Any movement away from it sabotages the whole concept of natural selection and Darwin's idea.

The Cambrian Explosion

Things lasted like this for about another two and a half billion years. Two thousand four hundred million to be more exact. Quite a long time really when the only living things on the planet were the tiny single-celled bacteria and algae in the oceans. Everything was alright as it was. There wasn't any sign of a struggle for survival leading to natural selection. Conditions apparently were just right for everybody. As I said – we're talking about a long time, a very long time. Try imagining – say – ten thousand years. Then fifty. Then a hundred. Then a million years. If you're like me, we'll be well outside our range already. But a thousand million – a billion – well... Anyway, it was a long time for nothing evolutionary to happen at all.

Then something did happen. About only six hundred million years ago. Suddenly and strangely – from any point of view, particularly the Darwinian – lots of weird and wonderful animals started to appear – fully formed – in the sea. From nowhere. And not so much as a by your leave. The fossil record shows it. Nothing leads up to them. There are no intermediaries. One 'moment' they're not – next 'moment' they're there. Now you don't see them – now you do. And many of them are still with us today: sponges, jellyfish, trilobites, sea-cucumbers, brachiopods and a lot of others, including a variety of sea-snails, sea-urchins and starfish.

Now these are pretty complex creatures. Their sudden arrival is a mystery that cannot be explained at all by the crude concept which is the basis of Darwin's theory.

Significantly, this phenomenon is called 'The Cambrian Explosion', for that is an honest and accurate description of this strange event in the history of life on Earth. It is as much an explosion as the 'Big Bang' which inaugurated the visible Universe. And equally unexplainable.

There are two points to note here. If Darwinians are right, how can one explain the sudden bursting in on the scene of the early Cambrian creatures? There is no gradualism here as demanded by the selection theory. And secondly, and conversely, why the incredibly long wait of two and a half billion years during which nothing happened – a vast period in which the mechanism of the theory was inoperative? If it is worth anything, why was it totally inactive?

The same thing happened a mere hundred million

years later. One could call it the Second Cambrian Explosion or The Cambrian Explosion: Part Two.

The earlier Cambrian creatures were all invertebrates: sponges, jellyfish, and others. In the second Cambrian stage vertebrates suddenly appear, in particular those wonderful animals, fishes. There's a bit of a difference between a fish and any of the creatures just named – but no fossil intermediate between them and a fish has ever been found. The fish just swam on to the scene with all fins a-firing and both gills a-flapping without so much as the slightest introduction or 'by your leave'.

Such abruptness is somewhat abrasive to Darwinists – to put it mildly.

Now fish are rather wonderful creatures. Though all creatures are if you think about them, down to the so-called simplest – the bacteria. Though there's nothing simple about them – as we have seen.

But fish. Well, they've always fascinated me. They can actually breathe air under water! Through those extraordinary things called gills. Which actually extract oxygen that you can't see, that you wouldn't even know was there, oxygen dissolved in the water. If you didn't know they could do it, you wouldn't think it was possible. But they do do it – and no one thinks any more about it. Ah, the taking-for-grantedness of man! Especially that curious species of man who call themselves 'biologists' who, when faced with such miracles – or at the very least wonders – simply nod and say with a knowing wink: "Ah yes, fish 'evolved' to fill and 'exploit' the evolutionary niche or space available."

In other words, there was nothing odd about it. It was bound to happen – sooner or later. There was notice for a niche in the environment marked 'Vacant' – and it had to be filled. Or, as they like to put it, "An evolutionary opportunity had to be taken." Very nice and pat. And how? Through the shuffling about of natural selection and a fooling about with the fittest.

Getting back to fish themselves, there are a lot more interesting things about them. Not only can they see, and see very well, but they've got ears, though you may not recognise them, and can hear. They also have a sense organ called the 'lateral line' which enables them to feel currents and other movements in the water around them to guide their swimming. I could go on and on. I used to keep them. They know a lot more than we like to think. Just because they can't talk about it.

As has been said not only fish, but many other creatures arrive all at once in the Cambrian period. Again, there are two points to observe. One is the sheer complexity of these beings – inexplicable from the orthodox evolutionary point of view. The other is their diversity – which means the mysterious arrival of sheer difference. Not only did most phyla appear virtually together at this time, the major groups into which naturalists like to place animals or plants, but within these phyla the 'classes' into which they often artificially place creatures which are utterly different from each other. The first representatives of all these groups were already so highly differentiated and isolated at their first appearance that none of them can be considered even in the remotest sense as intermediate with regard to other groups. Thus,

Darwin's theory of a common 'ancestor' to the different species breaks down through the sheer suddenness of the arrival of difference. As time itself and gradualism cannot explain this matter we are forced to consider the possibility of an extratemporal source of operation. Provided, of course, that our imagination is not totally atrophied.

Let us now think about the next and major event in the story – the arrival of terrestrial or semi-terrestrial animals (the amphibians). The question is: why, in standard selectionist Darwinian terms, should any creature ever have ventured out of its aquatic habitat – the medium in which it was so successfully at home for so many millions of years? Why leave an environment for which it was obviously so well adapted – for an environment for which it was totally unadapted – which was utterly alien, adverse and unsuitable for it in every possible way? What gain, what survival advantage could clambering on to dry or even wet land possibly have for the aquatic animal?

An entirely new way of breathing would be necessary. Oxygen would have to be taken direct from the atmosphere, rather than dissolved in water. New organs – lungs – would have to replace gills. Legs would have to replace fins for moving on the land. Desiccation through exposure to the heat of the sun would have to be prevented. In every way the terrestrial environment was a no-go area. In order to survive in this totally different element the candidate for survival would already have to be fully equipped for this purpose. The process of natural selection and the survival of the fittest, etc., would not suffice for such a major change. No gradual intermediate steps would do. It had to be all or nothing. Otherwise the fish would lie gasping on the beach.

Minor changes due to mutations would render the aquatic creature unfit for water as much as for land. Thus, according to the theory, it would in fact be selected *against*. It would not be as efficient as its fellows in the sea. It would not be suitable for land or water. The theory would prevent rather than aid its development.

A fish would find no survival value in undergoing any so-called aquatic-terrestrial migration. It is obviously much safer in the sea – the medium it is adapted to, the environment where it belongs – its natural home. No 'selection pressure' can alter this. There is no point in any attempt to move to the land. No reason for it. No need for it. The fish is perfectly alright where it is, in the element which it is clearly designed for. To venture out of this environment would be dangerous, foolhardy and quite unnecessary – having no survival advantage whatsoever. Rather the opposite. The prerequisite for survival is existence in equilibrium with an environment a creature is adapted to. Not a totally unnecessary sortie into a hostile world of absolute vulnerability. There is clearly an impasse here (and, as we shall see, elsewhere) which cannot be crossed in terms of Darwinian 'evolution'.

So any hypothetical struggle on to the land out of the sea would not be a struggle for survival, but a struggle *against* survival.

The same principle applies elsewhere. In fact virtually throughout the entire evolutionary 'sequence' – which, it is argued, is not a sequence at all but a series, or better still, an order.

The differences between a fish and an amphibian, an amphibian and a reptile, a reptile and a mammal or a bird cannot plausibly be explained away by chance winnowing and adaptation to the environment. They are too great. The beings concerned are too distinct, so wholly different, so perfect in themselves – as we have said before, and shall doubtless say again so singularly significant is the point, that they can only be envisaged as the products of nothing less than a holistic 'blueprint', a total conception. Their unlikeness is uncanny. Both their difference and their superb selfness declare design.

The Plants

We've mentioned the blue-green alga, but now it's time to look at those other wonderful beings, equally mysterious, who inhabit the kingdom of plants. The history of their 'evolution' is equally unexplainable as that of the animals. But for some reason it tends to be neglected by those who discuss this subject.

After the blue-green algae – it is not clear when – all kinds of larger algae, the waving and variously hued seaweeds, appear in the sea. Nor do we know how, because the seaweeds in all their variety are not just colonies or composites of the single-celled alga but distinct plants, beings, in their own right, with their own shapes and character, from the sea-lettuce to the bladderwrack – though in fact all plants in the sea belongs to the algae family, possessing the ability to breathe and nourish themselves through the entire surface of their bodies, by absorbing all the elements they need from the surrounding medium, the sea, and photosynthesising all over from every pore in every frond. This overall absorption of minerals dissolved in the water means that they don't have to have roots, which do this for land plants. And while they don't have true leaves as organs of respiration they can photosynthesise from all over their body.

Thus, the sea plants are totally adapted to their aquatic environment and in absolute equilibrium with it. There are two points here:

The first is that given that this self-sufficient and balanced relationship with its environment is true of 'even' the blue-green alga, why should it ever need to be any different? Why should it ever 'wish' to become a seaweed? In Darwinian terms, what selection pressure could there possibly be, what survival advantage could possibly be gained by turning into a seaweed? None at all. It was already in the best of possible worlds for a blue-green alga. No theory of natural selection would make it budge. No such notion could account for the arrival of the seaweeds.

The second point is of course why, by the same token, should the perfectly adapted seaweed 'wish' to leave the sea for the inhospitable and in every way unsuitable terrestrial environment – the land?

Have you ever seen a seaweed strolling onto the land? Even the bladderwrack, well supplied with little 'pockets' containing water will soon dry and die unless rescued by the return of the tide fairly soon. You won't see them on the promenade!

However, the accepted theory is that somehow or other certain green algae 'moved' up the rivers gradually adapting themselves to fresh water – and thence pulled themselves on to the land – provided there was plenty of moisture there. Then later other plants appear, like, and including, the hornworts, liverworts, and mosses, still very much with us today. They're called bryophytes, and can't grow more than a few inches high because they are non-vascular – that is, they lack the firm conducting veins which enable the vascular plants like cowslips or conifers to draw water and dissolved nutrients up their stems. But we'll return to this later. And with the same recurring question – how did such an extraordinary 'change' occur? For over and over again strange things happen – extraordinary things – which cannot be glibly explained by the word 'evolved'.

So why should the seawater alga enter the world of freshwater anyway? Why push upstream up the rivers into the, for them, inhospitable environment of ever decreasing salinity and alteration of mineral content? There was no survival advantage in this. Rather and obviously the opposite. It is all going against the grain.

Now there is a big difference between the alga and the bryophytes. Nor are there any transitional between them as required by the theory.

In size and shapes the bryophytes are totally different to the tiny algae who are held to be the first 'colonisers' of the land. They are also capable of withstanding desiccation by going 'on hold' for long periods when terrestrial conditions were dry. Furthermore, bryophytes, unlike the algae, produce spores for reproduction equipped with a hard protective coating in order to withstand dehydration. The hornwort even has stomata with guard cells for the control of loss of water through transpiration – a total innovation, characteristic of higher plants.

And what is the significance of the fact that they – mosses, liverworts, and hornworts – are still very much with us today? That bit of moss over there has lived on the planet for over 400 million years. Why hasn't it changed to fit in with evolutionary theory as it is supposed to? Even a very little? It shows absolutely no sign of being and becoming anything other than itself.

The bryophytes remain, and have for half a billion years remained, as they are – perfectly adapted to their habitats, in equilibrium with their environments perfectly content not to have a vascular system – which they do not need for their way of life.

The 'vascular' plants are those which have a fluid-conducting system made of xylem and phloem tissues which enable water and dissolved nutrients (minerals) to flow up from the roots to the rest of the plant. This allows plants to grow taller, providing the structure and food-distributing system necessary for such growth – ultimately manifesting in trees. Without this, water cannot be carried more than about half an inch upwards – which naturally restricts the height of the bryophytes.

The arrival of the plants with a vascular system is another mystery. This is undoubtedly a major advance – but again, there are no intermediates to provide evidence of any gradual evolution from the mosses, liverworts, and hornworts.

For millions of years the bryophytes had remained as they were, the only plants other than the algae to inhabit the terrestrial regions of the planet, perfectly adapted for what they wanted to do, without a trace of any incipient water-conducting vessels, until suddenly we see the arrival of the vascular plants imprinted in the fossil record. And interestingly and significantly they continued to exist alongside the first vascular plants and do so to this day. Far from 'evolving' into vascularity, they diversified into a host of different species horizontally, as it were. Today there are about 6,000 species of liverworts and about 10,000 mosses.

However, with the arrival of the water-conducting plants other changes were to take place, both unnecessary to, and unexplainable by, crude Darwinism. Roots appeared, allowing plants to extract and absorb water from much lower into the soil than the bryophytes were able to do with their rhizoids – which only enable the plant to fix itself to the ground like the hold-fasts of seaweeds. So plants with roots could colonise dryer soils inaccessible to the liverworts, mosses, and hornworts which needed damp environments for their survival.

The arrival of the root is an extraordinary thing. It is an entirely new organ. It is a specialised structure enabling the plant to absorb water and minerals dissolved in it from the surrounding medium in the quantities necessary to, ultimately via the vascular conducting system, supply the whole plant.

Then followed leaves. Even more extraordinary. No apology will be given for using that word so often. It is the correct term for the mysterious appearance of these entities one after another. The right word is probably 'miracle'. But I do not wish to be too dramatic for some of my readers. Nevertheless, if something occurs which is unexplainable in terms of or from what is present within a given situation, then it follows that its origin is outside that situation – and if also there is an element of design inherent in what is being referred to, it is legitimate to use that term.

The leaf is a lovely thing. And the leaf of each plant is so delightfully different, each with its own design, its own character. I look out of my window and see them – in all their various and wonderful shapes. One has to be perversely purblind not to see that such patterns cannot be the product of chance. Just look at those intricate vein tapestries, as complex as arabesques.

But of course the leaf also has its special role in the life of the plant. It is the site of photosynthesis, which takes place in the hundreds of chloroplasts open to the sunlight – or closed or partly closed by the action of the stomata and their guard cells which control the flow of transpiration through the plant, and regulate the inflow of carbon dioxide from, and release of oxygen to, the atmosphere. No one knows how.

There is design too in the manner in which leaves are placed and arranged around the stem of a plant. In essence leaves form a helix pattern centred around the stem, either clockwise or anticlockwise, with – depending on the species – the same angle of divergence. It has been discovered that there is a definite regularity in these angles and they follow the numbers in a Fibonacci sequence: 1/2, 2/3, 3/5, 5/8, 8/13, 13/21, 21/34, 34/55, 55/89. This series tends to a limit close to $360°$ $34/89 = 137$ 52 or $137°30$, an angle known mathematically as the golden angle. In the series the numerator indicates the number of leaves in the arrangement. This can be demonstrated by the following:

> Alternate leaves have an angle of 180° (or ½)
>
> 120° (or $^1/3$): three leaves in one circle
>
> 144° (or $^2/5$): five leaves in two gyres
>
> 135° (or $^3/8$): eight leaves in three gyres

Another inexplicable occurrence is the arrival of the seed, and the seed plants. The bryophytes used spores for reproduction, and the spore itself is undoubtedly a remarkable thing. But the seed is even more remarkable; it is a vessel which contains within itself no less than a complete embryo of the particular plant. Together with a supply of food to nourish the embryo when it begins to grow, well into the time when it has developed its first roots – in order to give it a good start before it can fend for itself from the surrounding soil. The seed is something which should not be taken for granted. It arrived fully formed – like so much else in the world of life. There were no intermediate stages by which it can be said to have 'evolved'.

The Flowering Plants

The mysterious arrival and origin of the flowering plants has puzzled orthodox evolutionists ever since it puzzled the producer of the theory himself. It so defeated and depressed Charles Darwin that he called it "an abominable mystery" for he was as aware, as we continue to be today, that the fully developed flowering plants suddenly irrupt on to the scene in considerable profusion of kinds about 140 million years ago in the period known as the Cretaceous –

without a trace of evidence in the fossil record of anything leading up to them. Despite another 150 years of searching for intermediates between them and the non-flowering plants like the ferns, the Conifers, Horsetails and Quillworts, the position remains the same today.

D.H. Scott, a distinguished botanist, observed in his work *The Evolution of Plants*:

> "...the apparently sudden appearance of quite well-developed Flowering Plants is still, perhaps, the greatest difficulty in the record of evolution."

Axelrod, another expert in the field, concurs. In his book *The Evolution of Flowering Plants* he admits:

> "...there is still as much of a mystery today as Darwin encountered surrounding the early evolution of the flowering plants notably their centre of origin, their ancestry, and their sudden appearance in the Cretaceous as a fully evolved, wholly modern phylum... The ancestral group that gave rise to these plants has not been identified in the fossil record, and no living representative points to such an ancestral alliance. In addition the record has shed almost no light on relations between taxa at ordinal and family level."

And not only their arrival – for that appears to be the right word rather than development – but the sheer diversity in which the flowering plants almost simultaneously appeared, adds to the mystery, if indeed it needs to be added to. Why so many? Wouldn't *one* do? Instead of a single or even a few 'vertical' breakthroughs we are presented with a vast horizontal array of thousands of flowering species.

It is almost as though a cosmic seed-scatterer had sowed a very mixed 'packet' of seeds on a grand scale.

Heribert-Nilsson noted in his extensive study of plant fossils that fossil evidence is lacking to support the evolution of *any* plant group. He even concluded that the fossil record concisely indicates that plants did not evolve but "flared up" in a non-evolutionary manner.

It is interesting to see the conclusions reached by Dr. E.J.H. Corner, a leading expert of the Department of Botany at the University of Cambridge – all the more so because Corner is still an evolutionist, albeit an uneasy one.

> "Evidence can be adduced in favour of the theory of evolution – from biogeography and palaeontology, but I still think that, to the unprejudiced, the fossil record of plants is in favour of special creation... The evolutionist must be prepared with an answer, but I think that most attempts to answer would break down before an inquisition."

That is an extraordinary admission from someone who nevertheless still likes to regard himself as a committed Darwinian. Which is of considerable psychological interest in itself.

But the reason for such a statement has to do with the fact that the fossil record consistently shows, in the words of Leclercq:

> "...persistence of type with imperceptible change and, from time to time, the sudden influx of new types, correlative with favourable stable geological conditions, are among the outstanding features of the history of evolution as shown by palaeontology."

This observation of course equally applies to the animal kingdom – as we have attempted to show earlier.

The lack of fossils is not the real the problem for evolutionists. But rather the problem is the lack of evidence for evolution in the *abundant* record that now exists which shows clear evidence for stasis rather than dynamic algae to flowering plant evolution.[1]

---

[1] I am fully aware that the faculty of choice I appear to have credited creatures with as to their preferred environment is not strictly Darwinian theory. But I would argue that it is a valid way of expressing and emphasizing the much greater unlikelihood of entry into a new and quite different environment taking place merely as a result of chance variation and survival pressure. It is simply a device for throwing the absurdity of the latter into relief.

CHAPTER 4

THE SIGNIFICANCE OF

SYMBIOSIS

Though there are other more diluted definitions of 'symbiosis', the essential and original meaning is 'living together in a mutually complementary fashion'. This occurs when two beings or species co-exist with each other in a mutually beneficial relationship, each giving the other something that it needs, in a productive inter-dependence. It is not a phenomenon easily explainable for Darwinians for the obvious reason that it runs counter to their central tenet of an ongoing competition and struggle for existence in nature, the assumption upon which the theory of natural selection is based.

The classic example of symbiosis, and also the perfect symbol for it, is the lichen. I can see one now growing on the slate roof opposite me. I say 'one', and we call it a lichen, but strictly speaking it is two beings closely intertwined with each other, an alga and a fungus, knitted together so closely for mutual support that we regard it as a single plant. But really there are two. The alga photosynthesises, giving the fungus the organic matter it needs for its survival but can't produce for itself, being unable to photosynthesise. In return, the fungus provides its partner the alga with a protective covering against the desiccation which otherwise it would be vulnerable to. It also helps the alga to absorb certain inorganic nutrients, particularly phosphorus. Thus, lichens are extremely tolerant of almost total desiccation, and can colonise – as we can see – exposed bare areas where no other plants are able to survive. So close is the association of the two plants that the alga actually lives *within* the tissues of the fungus, which is adapted to this inhabitation.

The lichen alga and fungus are not found apart, growing on their own, in nature.

There are many interesting examples of symbiosis in the natural world. One is the close relationship between the tiny floating fern known as Azolla and the blue-green alga called Anabaena. The alga is a nitrogen-fixing species able to take nitrogen as a gas straight from the atmosphere and convert it into a form acceptable to other plants – nitrates. The floating fern thus gets the nitrate nutrient it needs from the alga, which actually lives inside its leaves. In return it provides the alga with secure accommodation and certain organic nutrients.

A very common symbiotic relationship is that of the legume family with the nitrogen-fixing bacteria Rhizobium. This is a very important and widespread family which includes not just the bean and pea, but clovers, vetches, lupins, gorse, broom, laburnum, the acacia and many others. This relationship plays a major role in the entry of nitrogen into the ecosystems of the biosphere generally, and much use of it has been made in agriculture. For example wattles, which are acacias, are crucial to the Australian ecosystem in their role as capturers and transmitters of nitrogen. And the semi-deserts of Africa rely upon acacia trees and their partnership with Rhizobium in exactly the same way. The same can be said of the role of clovers on the hills of Ethiopia. So we can see that this symbiosis benefits not only the immediate members involved, but plays an essential part in the survival of the whole natural community in some cases.

So close is this relationship that the bacteria concerned actually live in tiny nodules in the roots of the leguminous plants. It is through these nodules that the roots can absorb the nitrate nutrients which Rhizobium makes available to them.

Furthermore, neither the Rhizobium, by itself, nor of course the legume can extract or fix nitrogen directly from the air. Only together can they do it – in a close and complex relationship in the plant's nodules.

In fact there are about 250 species of plants other than legumes which also form symbiotic relationships with nitrogen-fixing bacteria. Significantly, they live in areas where soil nitrogen is low. The alder tree and the bog myrtle have nitrogen-fixing bacteria known as Frankia in nodules in their roots.

We must add, though, that almost as fast as these bacteria are producing nitrates from the nitrogen gas in the atmosphere, other bacteria are busy breaking it down again and releasing it back into the air. So plants have to work quickly to absorb their share. It clearly pays to do as the legumes, the bog myrtle and the alder do – keep the nitrogen-fixing bacteria on the premises!

But it is important to realise that not only the legume itself and its bacterial partner benefit from their symbiosis – but the overall fertility of the soil is increased. A striking example of the capacity of nitrogen-fixing bacteria to improve the fertility of the soil was seen in an experiment carried out by the United States Forest Service near Athens, Ohio. A planting of Cedar trees was set out in an area of very poor soil. In one part of the area a number of locust trees were set out among the Cedars. Now, locust trees are legumes, carrying nitrogen-fixing bacteria on their roots.

Eleven years later, the Cedar trees that had been planted alone averaged 30 inches high, while those planted among the locust trees had grown to an average of 7 feet.

There are about 13,000 species of legumes on the planet that, with their bacteria, fix 100 million tons of nitrogen every year. This is obviously an occurrence of significance for the biosphere. Without this constant enrichment, Earth's soil would soon become too poor to sustain the quality and variety of plants that we now witness, and all the ecosystems that depend on them.

Let us now look at cases of symbiosis between plants and animals. A very interesting example is the cooperation that exists between a Central American Acacia and a particular ant of the genus Pseudomyrmex. This Acacia has hollow thorns, and pores at the bases of its leaves that secrete nectar. Now, these hollow thorns are the exclusive nest-site of the ants who also drink the nectar. The Acacia provides them with both food and board! However, in return, the ants protect the plant by driving away plant-eating insects and prune back vines and shrubbery that might crowd out the Acacia.

All around us, of course – and without which nature would be much the poorer – is a symbiosis of great importance which we hardly notice. That of insects, particularly the bee, and the flowering plants. We all know of, but may not have consciously recognised as such, the symbiotic relationship between the bees and the flowers which they pollinate thus enabling the plant to reproduce, and getting nutrient-rich nectar in return.

In temperate countries like ours the bees tend to be versatile. They will feed on all the flowers of the meadow, or on heather or clover, depending on what is available. But in tropical forests many species of plants are pollinated only by particular species of insect. Thus, in Amazonia there are at least 80 species of fig, each pollinated by its own species of wasp. One eighth of all flowering plants are orchids – 30,000 species out of 250,000 – and many of them, in tropical forests, are pollinated by their own particular insects.

Such extreme specialisation, such 'putting one's eggs into one basket' runs counter to the requirements for survival inherent in the Darwinian model – as indeed all symbiosis does.

It is worth noting that the bee is extremely well equipped by 'nature' for its role of pollination by which both plant and bee benefit. Through its antennae, a bee can detect scents 100 times more dilute than those we humans can smell. The bee also has a hairy tongue specially adapted to the absorption of nectar, and different species have tongues of different lengths, enabling them to extract nectar from a range of flowers of many different types. The symbiosis is thus acutely dovetailed.

While we are on the subject, let us look at the association between an orchid of Madagascar and a moth with which Darwin's name has long been connected. This orchid has a foot-long tubular nectary, and when Darwin first looked at it he knew from experience that these plants usually have a single insect pollinator. But to reach the inch of nectar at the bottom of this orchid's nectary the insect concerned would need to possess an incredibly long proboscis. So Darwin made a bold prediction that there exists in Madagascar an insect with a proboscis twelve inches long. Entomologists scoffed at the idea of such an insect. The whole thing was preposterous. But the scoffers were silenced when, several years later, a previously unknown Madagascan moth was discovered – foot-long proboscis and all. Subsequently, as it happens, the story repeated itself in reverse, when in South America a moth with a twelve-inch proboscis was first discovered. Then after a considerable time, a corresponding flower with a foot-long nectary was found.

Darwin's forecast on this occasion is usually regarded as an indication of his 'genius'. For many it shows how much he knew what he was talking about. However, a hard look at the incident actually indicates the very opposite.

His prediction here was in fact anything but based on his theory. It was simply that of a naturalist – the very good naturalist he undoubtedly was. It was merely an extrapolation consistent with his own past knowledge – of the symbiosis of orchids and insects. It was a reasonable common-sense deduction from the known facts of nature as he found it, and as we all find it. Not genius. In fact, it runs counter to his theory of natural selection. As symbiosis does in general. Here we have two creatures in the precarious position of totally depending on each other – a vulnerability which is the very antithesis of fitness for survival. The theory should predict that it should *not* happen. So Darwin's prediction was not a scientific prediction based on his theory. It went directly against it. But as the prediction of an observant student of nature it was of course valid. So the irony is that Darwin has been praised for something he accurately guessed, which contradicts the very premises he was advocating. Whether he saw this or not, is another matter.

Darwin's theory would require the flower and its pollinating agent to evolve in a parallel and complementary manner, which is improbable in the extreme. The first flowers would need pollinating agents in order to survive. And vice-versa. The mutuality is a must from the start. One is a sine qua non of the other. It is a both or nothing situation. We are witnessing in fact a *system*. A co-ordinated interaction. And systems don't arise by chance.

There are many more examples of symbiosis one could mention. Cows need certain bacteria within their stomachs in order to break down the cellulose which is so large a part of their natural diet. So do we need the bacteria in our digestive tract in order to have certain essential vitamins, for example biotin, which they produce for us.

A particularly striking symbiotic pollinating relationship is that of the hummingbird and the flowers it feeds upon. This small and delightful bird can, as we know, hover in mid-air, beating its wings extremely rapidly up to 90 times a second, which enables it to hold its position in front of the flower from which it extracts the nectar essential for its livelihood. It is also the only bird that can actually fly backwards – which is another useful faculty in its flower-visiting and pollinating way of life.

The hummingbirds' nectar requirements are extreme in order to maintain the incredible rate of their wing-beat. They have to consume more than their own weight of nectar each day, and in order to do so they must visit hundreds of flowers daily. With the possible exception of certain insects, the hummingbirds when in flight have the highest metabolism of all animals. This puts them in a vulnerable position. Hummingbirds are in fact continuously only hours away from starving to death, and are able to store just enough nectar to give them energy to survive overnight.

Hummingbirds are specialised nectarivores and are tied to the flowers they feed from. Some species, especially those with unusual bill shapes such as the Sword-billed Hummingbird and the Sicklebills, are adapted to be linked with only a small number of species of flowers. Most hummingbirds have bills that are long and straight or nearly so, but in some species the shape of the bill is closely adapted for specialised feeding. Thus, Thornbills have short, sharp bills adapted for extracting nectar from flowers with short corollas. The Sicklebills have extremely decurved bills which are adapted to getting nectar from the curved corollas of flowers in the particular family called Gesneriaceae. The Awlbill has a bill with an upturned tip suitable for feeding from another type of flower. And there are others whose bill shapes equally match the accessibility of the flowers they obtain their nectar from.

Such facts about the hummingbird speak for themselves. How can evolutionary theory possibly explain the existence of a creature with such a dangerously vulnerable and specialised mode of life. Adapted to survival it may be, but only within very intricately particular parameters which in fact make it not generally adapted for survival at all from a wider point of view.

Unlike other birds the hummingbird is committed to a need for energy from a narrow niche which renders it constantly at risk unless the extraordinary exertions required by that need are continuously maintained.

The symbiosis of bird and plant is obvious. Less obvious is how such an unusual and specialised relationship could have been arrived at by chance and survival pressure when it clearly runs counter to the very concept of survivability. But thank goodness the hummingbird exists, all the same!

We have looked at a number of particular examples of symbiosis, but there is a symbiosis so general in the biosphere, in which all living things are involved, and without which there would not be a biosphere, so ubiquitous and taken for granted that we tend to be unaware of it. Without it, I wouldn't be sitting here writing this book, and you wouldn't be sitting there reading it.

It's the planetary exchange of gases between plants and animals and humans necessary for their mutual subsistence: the production of oxygen from the plant kingdom in the course of their photosynthesis which animals and humans need from the atmosphere as part of their nutrition which would soon be exhausted without continual replenishment from green plants in return for the carbon dioxide breathed into the air by the animal and human population.

Without this perfect cycle life on Earth would have ceased a long time ago. Carbon dioxide is in fact a rare gas on our planet – much rarer than we might realise from the current focus on its role in possible climate change. It actually constitutes only $35/1,000$ of the atmosphere, even less than argon. That amount of carbon dioxide, if not replenished, would support the present plant population of the planet for only forty years. Thus, the respiration of animals and of certain bacteria is crucial for the continued life of plants. And without plants no animals would survive. Nor humans.

We have, on the whole, been looking at symbiosis from a strict point of view. We have identified clear examples where particular beings need each other for their existence on Earth, in particular ways.

But the more we study the great web of life on this planet, the more we begin to realise what an intricately interrelated complexity it is, evincing a vast pattern of interdependence, a mosaic of mutuality between all its creatures – bacteria, fungi, plants, animals and man – a symbiosis on a truly grand scale, an interwoven, pulsating, organic whole which surely cannot be the result of the blind workings of chance.

CHAPTER 5

THE MYSTERY OF
ADAPTATION

If there is anything that strikes one as one looks around at nature on this planet, it is the extraordinary way in which its creatures – both plants and animals – are adapted to the environments in which they live. This adaptation is often so specialised, so exact, such a perfect match for the surrounding medium as to be uncanny. To believe that it occurred by the chance survival of beings fortuitously gifted with the characteristics necessary to cope with their environments is indeed an act of faith – or folly.

We have already seen in our study of symbiosis a number of cases of adaptation of great and often curious exactitude. Let us now look at a variety of other interesting examples which will illustrate the opening remark of this chapter.

In fact every being on the planet exhibits a complex and exact adaptedness to the medium in which it lives, the more we look at it.

Look at the birds again. Their wings are wonderful things. We have already noticed how the sophisticated and intricate arrangement of flight feathers on each wing form the basis of an aerofoil of variable geometry. Which gives a bird the ability to vary the shape and aerodynamic properties of its wing at take-off, landing, and for various sorts of flight: flapping, gliding or soaring. It is perfectly made for the air.

Look at the fish. How finely fashioned is its finnage which enables it to maintain its poise of perfect relationship with the surrounding water, whether swimming through it, or holding itself still. And its gills – those extraordinary organs for extracting dissolved oxygen from the water which if we didn't witness the fish doing it, we wouldn't imagine it was possible.

Then the bat. Unique as the only flying mammal, and having not changed in millions of years, it is of course also unique in the extraordinary faculty of echolocation it possesses for living and flying in the dark. Bats produce a series of short pulses of high-frequency sounds from their mouths and when this sound strikes an object it is reflected as an echo. From these echoes a bat can determine both the distance and direction of any object in front of and at either side of it, rapidly building up a picture of its immediate environment by this remarkable feedback system. This enables it to accurately manoeuvre at speed in a dark cave without colliding with its walls. Not only this but the echolocation system is so developed that if the bat is hunting for moths and other insects at night, it can determine in addition to their distance and direction the speed, size, shape and surface texture of these insects as they fly.

Equally remarkably, certain moth families contain species which are able to detect marauding bats and take evasive action by using an organ they possess actually capable of generating a 'jamming' signal which disturbs the bat's echo patterns and confuses its hunting technique. How did the moth not only become cognisant of the bat's sonic intelligence but also 'evolve' such a defence against it? Anyone who says 'chance' must be joking!

As for the bat itself, the echolocation apparatus must be perfect to be of any use whatsoever. Both emitter and receiver must co-operate together in perfect co-ordination to be of service to the bat in its survival. Nothing less will do. Needless to say no intermediate fossil forms have ever been found in the fossil record.

The world of insects contains a number of examples of distinctive adaptation, but particularly unusual is the way of life of the water spider. This is the only spider which lives underwater, inhabiting freshwater ponds and slow-moving stretches of water here and there. Fascinatingly, this spider, which is about half an inch in size, creates a little bell-shaped nest filled with air attached to water plants under the water. It does this by repeatedly swimming (it can swim very well) up to the surface and back bringing down with it small bubbles of air attached to its furry abdomen which it uses to build up its hideaway which it then inhabits, breathing the air contained within this diving-bell. There it sits, to emerge occasionally to catch aquatic midges and other small insects.

Water spiders can remain submerged in their air bell for fairly long periods, but from time to time they must swim up again to the surface to replenish the air. However, frequent replenishment is not necessary because the structure of the bell, which is made from fine silk produced by the spider, is permeable to the air dissolved in the surrounding water, and thus permits gas-exchange with the aquatic environment through osmosis. Oxygen is replenished and carbon dioxide expelled due to differences in the osmotic pressure.

This system has been called "the water spider's aqualung". However, it is actually far more advanced than the real Aqualung which needs to be refilled frequently – not having the option of osmotic exchange. Not too bad. But wait for it – it has recently been discovered that the water spider can actively monitor the quality of the air in its bell. An experiment was performed in which the air volume in a water spider's bell was artificially replaced with either pure oxygen, pure carbon dioxide, or a control of ambient air. The spider only reacted to the carbon dioxide treatment, surfacing more often and ceasing its bell-building activity until the oxygen level had been sufficiently replenished. As it did not react to the oxygen or normal air situation it appears that the water spider is able to detect an increase in the proportion of carbon dioxide present in its bell over oxygen and act accordingly to restore a proper balance – critical for its survival.

Now, a question arises.

The spider is quite able to live on land. It is an air-breathing animal and can spin a web of some kind capable of trapping terrestrial insects. Why then has it 'chosen' to enter, or more significantly re-enter water – the aquatic environment from which life is supposed to have emerged? Why is it going back on evolution? Why is it making things difficult for itself – invading a medium utterly alien to its method of breathing, necessitating the construction of a breathing chamber requiring considerable craftsmanship and constant monitoring for safety? And also requiring a new mode of movement – swimming? How can natural selection possibly account for the process whereby the water spider gradually adapted itself to an aquatic existence? Exchanging an environment suitable and safe for a terrestrial creature for one alien and dangerous? One conducive to survival for one unnatural to it in every way. In fact an environment diametrically opposed to a normal spider's way of life. By what steps – according to orthodox selection theory – did it make a journey each stage of which was fraught with factors antithetical to its survival from terra firma to watery depths? The question remains. The answer is absent.

How too can accident explain the water spider's extraordinary ability – not merely to adapt to this – for an insect – unnatural medium, but actually consciously adapt its environment to itself by monitoring and altering the gaseous content of its underwater bell to suit its breathing requirements?

Let us now turn to the world of plants, the plant kingdom, for examples of adaptation of such sophistication that it cannot be explained by the primitive and crude method of natural selection.

We have already talked of the lichen. Its extraordinary adaptation to the inhospitable environments it so ably inhabits is in its union of alga and fungus. Without this symbiosis it would not succeed. Significantly the particular alga and particular fungus involved are not found on their own in nature. The lichen, then, is a dual whole whose design is inherent in its very existence and whose adaptiveness is inherent in its design. The design stares one in the face.

There is a strange plant to be found in the mangrove forests which grow in the salty waters of subtropical lagoons. All the plants in fact, of these forests, are specially adapted to living in salty water that would certainly be deadly to any other plant. Mangrove forests are best visited by boat, and surprise all who see them for the first time. Only the tops of the trees can be seen at high tide when the sea is flooding the forest. It is difficult to understand why trees should choose this as their normal environment. It is not something one would expect.

The tree in question is the *Rhizophora stylosa* tree. That cannot be its true name but it is the one it has been given by botanists. What makes it interesting is its peculiar method of reproduction, unknown to any other plant in the world and uncannily adapted to the environment in which it grows.

When the tree is in fruit sausage-like objects hang from the branches, up to two feet long. These objects are in fact the developing seeds, which germinate while the fruit is still on the mother tree. From the pear-like fruit a hypocotyl of the seed slowly grows, elongating and thickening. Finally, the sausage-like hypocotyls become too heavy for the branch to support them and like a dagger they plunge straight into the mud or water below. By striking the mud with some force the heavy seedling penetrates the mud and, by quickly taking root, is able to successfully withstand the returning tide. If, however, the tide is in and it lands on water, it floats on the surface and has a good chance to take root elsewhere when the tide is out again. Normal seeds would be carried away by the sea current and perish.

This adaptation of the Rhizophora tree is clearly 'tailor-made' to fit its existence in the mangrove forest. And there is also another adaptive peculiarity which it shares with the other members of the mangrove community in general. This is their unique ability to equip themselves with oxygen when they are submerged and unable to get it from the air, and also when they have to compensate for its deficiency in the salty mud. When the tide is out one can see this peculiarity of the mangrove forest very clearly. Pencil-like projections stick up out of the mud around the trees. They are in fact special breathing roots which are able to absorb oxygen from the atmosphere and retain it to supply the plant both then and for its sojourn under water.

To believe that such adaptations came about by chance one must have a peculiar kind of faith.

The Cactus

Another plant whose adaptation to its environment is extraordinarily sophisticated is the cactus. Inhabiting the dry regions of South America, Central and North America and also southern Africa, the essential problem for the cactus is how to attract and conserve water. And this it does in a masterly manner.

As most water loss takes place through the leaves of a plant by transpiration through the stomata, those tiny pores which release water vapour in the sunlight and simultaneously absorb carbon dioxide necessary for photosynthesis, and of course extract and emit oxygen, the cactus has overcome the problem of losing its limited supply of water in the environment in which it has to survive in a complex of different and for the most part unique ways.

Firstly, the cactus doesn't have leaves – through which most transpiration occurs. Instead it has, as we all know, spines. This means also that the surface of the plant exposed to the sun is greatly reduced. The spines not only shade the body of the cactus, from which transpiration also occurs to some extent, but at night help to condense and retain dew – a valuable source of water for this plant where rain is so rare. They also play the important role of protecting it from moisture-seeking animals roving the semi-desert landscape.

Secondly the stomata of the cactus are closed by day, closing the 'gateways' of transpiration when it is at its greatest and most dangerous to the life of the plant, and are opened at night when there is much less water loss in this way. A certain amount of transpiration is of course necessary for the plant to provide it with a flow of water and dissolved nutrients.

The stomata must also be open at night to absorb the carbon dioxide so necessary for photosynthesis for the cactus, like all plants, from the atmosphere. It stores this and releases it gradually during the day when photosynthesis occurs, as it does for every plant in the presence of sunlight. However, the so-called 'dark reactions' of the process of photosynthesis are performed by the cactus at night.

The whole point is that the cactus has neatly solved the problem of how to exist in an environment hostile to the existence of plants by virtually separating the processes of transpiration and photosynthesis – ensuring that they do not occur at the same time. The former happening by day, the latter by night.

But there is more to come.

Photosynthesis is carried out by the cactus through the whole of its body-stem, which is made of thick, water-retentive tissue and usually has the form of a sphere or cylinder – the optimal shape for combining the highest possible volume (particularly with storage of water in mind) with the lowest possible surface area. A full-grown Saguaro cactus can absorb and hold 3,000 litres of water in the short space of two days.

It also takes rapid advantage, as it has to, of any rainfall through its roots. All cacti have a wide spreading root system in order to take advantage of brief showers by intercepting the downward flow of water before it sinks too far for tapping. But a young Saguaro about four and a half inches tall has been found to have a ramified root system over six feet in diameter and with no roots less than four inches deep below the surface. And it can form roots very quickly. New roots appear only two hours after rain following a long draught, and an extensive root system spreads out immediately under the surface.

Finally the salt concentration in the roots of the Saguaro and many other cactuses is so high that when water is contacted it can immediately be absorbed in the greatest possible quantity.

The cactus therefore is peculiarly designed, or if one prefers, adapted to its dry environment not just in one but in a combined cluster of ways. So complex and conjoined are these characteristics that it is surely impossible to believe they all developed by chance. The cactus reveals as a whole being and in all its individual features what Blake would call 'the lineaments of design'. Adaptation on so sophisticated a scale implies an Adaptor. Design implies a Designer. We are not talking about belief or faith of any kind here for this to be accepted. Merely logic will do. And the logic is inescapable.

There are no traces of 'intermediates' for the Rhizophora tree of the Mangrove forest. Nor are there for the cactus.

Two general points can be made about adaptation in nature. Firstly, its ubiquity. Wherever we look, we see it. Animals and plants are everywhere perfectly adapted to their particular environments, no matter how diverse and often inhospitable. Almost every environment is inhabited – from the depths of the oceans to the dry semi-deserts. From the cold of icebergs to almost boiling water of hot springs. Water, land, and the air itself all have their inhabitants. Why? Why should it be that living creatures have entered and invaded areas of the planet where life would be impossible for them without very particular and extremely specialised adaptations? Why make things so difficult for themselves in the so-called 'struggle for survival'? There was, on the face of it, no need for it. Or was there, from an at present unknown point of view? We shall develop this later.

Adaptation does not *explain* adaptation. It does not simply take place because it is needed. And often from an ordinary point of view it is not needed. It is in no way necessary, on the face of it, for the creatures of Earth to be able to populate every part of the planet no matter how adverse a medium for survival it is. Why make survival even more of a struggle than it is, than it need be?

Secondly, adaptation is not the same thing as evolution. Nor does it explain it. The purpose, extent and limit of adaptation per se is only survival. That is its object, and that is its cut-off point. It cannot explain *development*. Adaptation is not advance. Advance cannot be explained by adaptation alone. Any more than adaptation can be explained by itself. Evolution cannot be explained by adaptation.

Yet the central Darwinian mechanism for evolution is nothing but chance arrival of variations with the survival capacity to be selected or 'winnowed' in preference to others by natural selection *for nothing more than survival capacity alone*. This is stasis rather than dynamics. There is no inherent forward or upward thrust in this situation. It cannot explain the kind of development we see in nature. Darwin's theory has, it is submitted, got too big for its boots.

Not only evolution but adaptation itself cannot be explained by trite orthodox notions. The kind of sophisticated and complex adaptation we have been looking at in, say, the cactus is in itself so phenomenal, so extraordinary, that it could not have come about merely through chance mutation and natural selection.

Selection is limited to whatever is available *for*

selection. And the *need for* survival doesn't explain the existence of the 'equipment' necessary for survival. We must be careful to notice the sleight of hand processes at work here, which commitment to Darwinism consciously or unconsciously produces.

The Peppered Moth

Nothing better illustrates the poverty or, more accurately the bankruptcy, of the evidence for the theory – if it can be called that – of evolution proposed by Mr. Charles Darwin than the case of the peppered moth. It has become the standard example for evolution through natural selection – mainly because it is the only example that has been found. Everywhere it is hailed as the clear and simple proof that Darwin was right, and regularly appears in the text books on the subject presented in schools and universities. However, even a cursory look at the 'story' reveals how shallow and obtuse is the thinking, if it can be called that, which purports to support this proposition, and has obfuscated the entire issue for those who have only a vague idea of what it is all about.

Let us look at it carefully.

In the early 1800s nearly all of the individual peppered moths (*Biston betularia*) were of a light-grey, speckled colour – hence their name. Active mostly at night, like most moths, they needed to hide by day when resting, from predatory birds. Since trees and rocks tended to be covered with mottled, light-green, grey lichens the moths were very effectively camouflaged.

There was, however, a rare variety of the moth

which was dark in colour which was therefore easily seen by birds and therefore prey to them. Which was probably why it was rare. About 98% of the species were of the light variety.

But as the newly industrialised parts of England became increasingly polluted in the course of the 19th century, smoke killed the lichens growing on the trees and blackened their bark. So of course the position was reversed. The pale-coloured moths which had been well camouflaged before became very conspicuous and fell victim to birds. And now of course the rare dark moths, which had been very conspicuous before, were now well camouflaged against the black background. Naturally the most common moth 'changed colour from pale to dark'. That is the phrase used by many Darwinians. A truly chameleonic achievement! Rapidly, the proportion of dark to light increased until 98% of the total were dark by the end of the 19th century.

This is certainly a good example of natural selection – understandably prized by Darwinians because in fact there are so very few examples available. But of course it is not an example of evolution – though that is also loudly claimed for it. The moth remains the same moth. A superior form has not emerged. All that has been demonstrated is the superiority of one kind of camouflage over another in a given environment. Hardly evolution.

But since the 1950s the situation has been reversed again. At that time the British became conscious of the need for the control of air pollution. Since smoke pollution began to decrease, the light-coloured moths (*Biston typica*) have started to become more common again. They have made a comeback in style and at the present time they constitute 98% of the population once more.

The pale and dark forms of the peppered moth are exactly similar in every way except their colour. They are simply two types of the same species, and of course interbreed. Inability to interbreed is taken as an indication of difference in species. Moreover both types, as we have said, existed before the industrial revolution. It is only the frequency of the different types which has changed. This is neither variation nor evolution. Nor is it even adaptation as a *new* phenomenon. The two kinds of adaptation already existed. It is merely a matter of natural selection *within* a choice of adaptations. It is in no way evidence of natural selection as an evolutionary process.

CHAPTER 6

EVOLUTION OR HIERARCHY?

If one asks someone 'in the street' whether they believe in Darwin's theory of evolution the answer tends to be yes – unless they are very religious. However, when one asks them what they actually understand by the theory it is usually vaguely limited to an acceptance that life originated somehow or other in the sea with lowly creatures and gradually over millions of years 'evolved' somehow or other through a succession of beings into finally man himself. They imagine Charles Darwin was not only the creator of this concept, but also showed how it happened and provided the necessary evidence – again somehow or other.

When one explains that the means whereby this evolution took place was something called 'natural selection' and describes this process to them as best one can, they begin to look doubtful. Generally speaking, what people mean by evolution is simply that life appears to have developed from lower forms to higher – and that *this* was essentially Darwin's idea. When they are informed that the general concept of evolution was not original to Darwin, and that it is natural selection that is the central tenet he produced, they are surprised and disappointed.

Naturally so, because the closer people look at the concept of natural selection, the more they tend to consider it unlikely to explain more than, at the most, the variation of species – not their origin. And even the variation of species is often difficult to account for in terms of the process of natural selection alone. Even the well-known case of the Galapagos finches requires a great deal of blind commitment to the theory, combined with a good sprinkling of albeit unconscious sleight of hand, to accept it. We will have a careful look at it later, but to the uncommitted eye and to the open mind – prerequisites for perception, which neither Darwin nor his followers possessed or possess to any noteworthy degree, those thirteen finches are so firmly and clearly themselves that their perfect differences cannot easily be explained away by the chance shuffle and stumble of natural selection.

When one is 'committed' to a theory one becomes the victim of that theory. To be committed is not necessarily a virtue – but literally a 'vice' in the sense of something that locks you in. One can be committed to prison. Which is the perspective one adopts in looking at something and which one consciously or unconsciously imposes upon the thing viewed. An interesting and relevant case of this is connected with the peppered moth. It is worth relating in full because it demonstrates the process at work so clearly.

As we have said, a dark variety of the moth, originally rare, began to increase its numbers in proportion to the light-coloured variety in the course of the industrial revolution in the north of England during the 19th and then the 20th century. The hypothesis that this was due to natural selection was first put forward by J.W. Tutt in 1896. It was a perfectly reasonable hypothesis, but it was not proved.

Nor did it have to be until in the 1950s when the concept of natural selection as a primary driving force in evolution was under attack by a number of biologists. The defenders of the theory were a group led by E.B. Ford at Oxford, one of whose members was Bernard Kettlewell. They concentrated on moths and butterflies and Bernard's experiments became their showcase example for "proving" the efficacy of natural selection. He chose to study the peppered moth. He firmly believed that this was a prime case of this process – and was determined to show he was right. An easy one, you would have thought!

Now Kettlewell knew that peppered moths are rarely found on tree trunks – despite previous assumptions – are never found in high concentrations, and like most moths never fly during the day. Nevertheless, in his experiment he delicately placed the moths on tree trunks, two or four to a tree, which is much higher than any normal concentration might be, and finally exposed them by day. The trees were in polluted forests and he placed equal numbers of dark and light moths on each one. The results were as you might guess, and exactly as Bernard hoped. All good stuff! The birds of course performed as expected and reduced the numbers of the now uncamouflaged light moths exactly as the 'experiment' demanded.

Or rather as the concept of natural selection demanded. It acted as a coercive agent upon Kettlewell, whose servant he became, losing all scientific honesty and objectivity under its influence. His belief so dominated him that he had to manufacture the evidence to support it. Craig Holdrege of The Nature Institute, himself a scientist, finds the whole episode significantly revealing with regard to science as a whole and its tendency to hold or be held by dogmas, either consciously or unconsciously. He discusses this in his paper 'The Tyranny of a Concept: The Case of the Peppered Moth'. He observes:

> 'Natural selection is still an idea that tyrannizes many minds...We see the rigidity of an idea that resists all attempts at modification despite a legion of facts weighing against it. What might have begun as a useful idea becomes despotic.'

Holdrege makes the point very eloquently. And he has identified something which can unwittingly occur in many spheres of life and thought.

Ironically, however, it is still almost certain that the original premise was right. This probably is a good example of natural selection after all. Which is a relief to the proponents of the theory because there are so very few examples. The rise and fall in the proportion of dark to light moths has it seems been due to which variety is better camouflaged against birds. But it is hardly a world-shattering thought!

Returning to the theme of this chapter entitled

'Evolution or Hierarchy?' I have said that it is not the concept of natural selection that most people think of, or actually find convincing when talking about Darwin, but rather the general concept of the continuous evolution of creatures from low to high over vast periods of time that they associate with his thesis in 'The Origin of Species'.

One can see why. When one looks at the range of beings as they present themselves to us, provided one is unaware of the glaring gaps and lack of intermediates in the fossil record, it is plausible to believe that one is witnessing a continuous succession of creatures one 'coming from' the other before it – a kind of historical parade of living beings. Provided also one doesn't ask what magic wand turned one creature into another, or who waved it. And when one describes natural selection, very few still believe *that* is the magic wand.

Thus, when one looks at the series: fish, amphibian, reptile and mammal, there is an obvious succession there of creatures emerging from the water to the land and which is borne out by the geological record too. Birds, however, present a group difficult to place. And while we might all agree that mammals are superior to reptiles, can we be so sure that fish are any less in intelligence than reptiles?

Nevertheless, it is understandable and undeniable that it is this general series that has led to the support of Darwin's theory of evolution by the majority of people – who have usually, as I have ascertained through many conversations on this subject, not really thought hard about his proposition that the whole process can be explained simply by natural selection working on the analogy of artificial selection. Needless to say, very few indeed have ever even attempted to read his book, which explains the ignorance on this point, and much of the supposed support for the Darwinism.

Though there is undeniably a series, and according to the geological evidence, a sequence of beings more or less discernible in the historical record of life on the planet, it is another thing to therefore claim or assume that this is a causal chain – one being springing from or emerging from another, the one before it, and so on. One does not have to go to Oxford to recognise the well-known fault in logic which this is an example of – namely 'post hoc ergo propter hoc' – 'after this, therefore because of this'. Sadly, upon such false premises whole edifices have been built. The colossus with the feet of clay for example.

Yet this is the optical illusion which leads many to accept the Darwinian assertion of continuous evolution, including Darwin himself who so wanted to believe it that he managed to suppress his own logical faculty and instead devote himself to extolling and elevating mere variation to paper over the crack which fairly ordinary thinking ability would reveal. Was he dishonest? If he was, he was, as has been suggested earlier, a victim of his own dishonesty. However, if one wishes to be kinder one can perhaps say that he was rather the victim of an obsession – an overriding obsession that dominated all his thinking – if it can be called that – an obsession that postured as a hypothesis and masqueraded as a theory.

In case I am thought to be too hard on Darwin and his followers, I have recently come across the critique presented by Samuel Haughton of this very notion, not long after the publication of his treatise on *The Origin of Species*. Haughton was Professor of Geology at Dublin University at that time, and a respected contributor to the debate. It is worth quoting his observations on this point in full:-

> "The most serious logical blunder committed by all who invent a theory of life from the geological succession is that 'Succession implies Causation.' It is argued that the Palaeozoic cephalopoda produced, in some way or other, the Red Sandstone fishes; that these in turn gave birth to the Liassic reptiles; that the non-placental mammals of the upper Oolite grew after some fashion, and ultimately produced the Tertiary mammals, some of which, in an unhappy hour, gave birth to man. The only fact at the basis of this astonishing inverted cone of reasoning is that these creatures *did* succeed each other in the manner described, and from this it follows, *post hoc ergo propter hoc*, that they succeeded each other in the way of cause and effect.

> I propose to test this strange theory by a corresponding theory of mineralogical succession of igneous rocks, which opens up a fertile field of speculation, hitherto unwrought.

> The igneous rocks of the Palaeozoic period contain abundance of feldspar, whose principal constituent is potash; the Mesozoic igneous rocks abound in soda, replacing potash; and in the tertiary period soda itself gives way to lime and magnesia. Viewed in the light of this philosophy here is a distinct indication that soda and lime are only allotropic conditions of potash. We may read the history of their formation in the crust of the globe – if we will only open our eyes and see it written.

If any chemist or mineralogist were to put forward such a theory of the origin of soda and lime as the foregoing, he would be regarded as a lunatic by other chemists and mineralogists.

How does it happen that a theory of the origin of species, which rests on the same basis, is accepted by multitudes of naturalists, as if it were a new Gospel? I believe it is because our naturalists, as a class, are untrained in the use of the logical faculties which they may charitably be supposed to possess in common with other men. No progress in natural science is possible as long as men will take their rude guesses at truth for facts, and substitute the fancies of their imagination for the sober rules of reasoning."

Haughton, as an expert geologist was acutely aware of those gaps in the fossil record, that lack of the necessary intermediates, which so fundamentally undermines the theory, and could see no reason to accept the glib assertion that unrecorded variations could connect what were obviously different creatures. And he rightly recognised the fallacy inherent in the form of 'thinking' – if it can be termed such – that led, and still leads, to the facile impression of an evolutionary continuity of species.

Nevertheless, the general concept of a hierarchy of living beings, a scala naturae, even the notion of the arrival of this hierarchy in chronological order, has been in the human consciousness for a long time, and a long time before Darwin. And, as has been suggested, it is this image that many have when they think of his theory of evolution and which predisposes them towards accepting it, or what they imagine to be it.

Plato, Aristotle and several other Greek thinkers had enunciated this concept in ancient times. Plotinus expressed a more subtle and sophisticated version of it. And it was inherited by the Middle Ages as the concept of the Great Chain of Being which was passed on to the Renaissance and continued to imbue later centuries with its essential vision, albeit perhaps unconsciously. It certainly influenced Erasmus Darwin, his grandfather, as can be seen in his book, and there can be no doubt that, through him, Darwin himself.

Parallel with, and sometimes fused with, the concept of the Chain of Being was of course the creation story of Genesis which clearly delineated the successive creation of creatures from the fish of the sea to human beings. The Bible was read and respected as a source of knowledge considerably more in the last five centuries than it is today, and the poetry and power of that first great Chapter resonated through the culture of the West and the consciousness of its peoples to an extent greater than can be easily realised.

These, then were the influences at work in the centuries prior to and including the 19th, and which have doubtless left their trace even on our present cultural consciousness, and, it is argued, is the underlying concept which is often mistaken for Darwin's theory.

The Great Chain of Being

According to the concept of the Great Chain of Being, everything in the Universe had its 'place' in a divinely planned hierarchical order, which was pictured as a chain vertically extended. Hierarchical in the sense of an order based on a series of higher and lower, strictly ranked, gradations or levels. "Place" in the hierarchy depended on the relative proportion of spirit and matter present in the thing or being concerned. The less spirit and the more matter – the lower, the more spirit the less matter – the higher the place in the chain.

At the bottom stood various kinds of metals and stones – regarded by our culture as inanimate objects, but perceived as in some sense alive and possessing some albeit low form of consciousness in the philosophy of the Chain. Higher up were various members of the plant kingdom – mosses, flowering plants and trees. Then come animals, ranked hierarchically from the ladybird to the leopard. Then human beings. Then angels of various levels. And finally at the top was God. He it was who not merely created the whole spectrum of beings but continually sustained them.

All creatures on the Chain were distinct, and all were created from above. It was a noble and beautiful vision of life which we have almost lost.

CHAPTER 7

THE DESIGN ARGUMENT

The view that there is design inherent and implied in the very texture of life as it manifests on planet Earth may not be scientific, nor does it *need* to be, in view of the narrow and unidimensional form of thinking which is meant by that term, but it is nevertheless, based upon a form of thought which has an authenticity of its own and is as reliable, and probably more so than the limited use of the mind which goes by the name of science.

It is in fact a mistake to want to claim that the design argument is scientific in order to establish its 'legitimacy', when its truth does not depend upon such a primitive criterion, of which it has no need, and whose limited terms of reference it cannot in any case fulfil – any more than, as it happens, the Darwinian theory of evolution can really be held to rest on a scientific basis when subject to close, even in fact less than close, scrutiny. We have already pointed this out in earlier chapters, but we will return to it later, identifying the main areas in which the theory fails to meet the scientific criteria which it purports to satisfy. In other words, it is strongly submitted that neither the design argument nor the Darwinian thesis is scientific, and this has to be recognised honestly by proponents of both views. The difference lies in the fact that whereas the Darwinian theory is presented as being, and depending on being, scientific, the concept of design has no need to rely on this kind of foundation.

Nevertheless, the argument for design is based on a perfectly good mode of reasoning, not only every bit as good as the mode of thinking adhered to by science but one which is adopted and used regularly in the process of scientific investigation itself, and without which very few of the discoveries arrived at by such means would have been made. And it is a feature of scientific thinking precisely because it is a valid feature of human thinking in general.

I am referring to common sense deduction and the drawing of inference. If one sees footprints in the sand it is reasonable to conclude someone has walked there. No one has ever actually *seen* gravity itself but it is reasonable to accept that such a force exists through inference from the effect it has on objects. That, of course, is how science itself arrived at this conclusion.

If one comes across an artificial object, like an umbrella or an urn, it is obvious that it has been made by a human being, and that it is not just the result of chance natural forces. If a man who had never before seen an umbrella sees one for the first time he will not have to possess a first-class degree in logic to decide it has been produced by a member of the human race. If he sees a complicated machine for the first time left at the side of a field, namely a tractor, he will not have to wait for the driver to appear and make it move to arrive at the conclusion that it was made by one of his own kind rather than by a series of chance events.

This was the central argument of William Paley, a priest of considerable intellect as well as faith, powerfully stated and developed in his work *Natural Theology* (1802), much admired by thinking people during and after his time. In 1876 Sir Leslie Stephen, who was in fact a humanist, said:

"His admirable lucidity and shrewd sense extort our admiration..." in his 'English Thought in the 18th Century', and more recently J.M. Keynes wrote in his 'Essays in Biography' 1933: 'Paley's writings are marked by a most noble lucidity, by a prosaic sanity.' Even Darwin was for a long time convinced by his argument; 'I do not think I hardly ever admired a book more than Paley's *Natural Theology*.'

Later of course he changed his mind. He became the slave of his own theory. Against his better judgement, one suspects. For it to be true he had to disown his earlier master. But it must have been a struggle.

In his *Natural Theology* Paley reasoned that a watch, which is obviously an intricate mechanism, and thus could not have come about by chance, clearly implies the existence of a watchmaker. More than implies – it is an inescapable inference, or, put another way, a logical deduction. So too, he reasoned, by analogy, a living creature which is infinitely more complex an entity, is bound to have a designer. Most of his book was a very thorough descriptive demonstration of the extraordinary complexity of various creatures which clearly indicated that they were the result of incredible design and in no way the flotsam and jetsam of chance.

The other central feature of the watchmaker argument which Paley employed, and which has become part of the general argument for design ever since, is that the watch has been made for a purpose – to tell the time, and thus it is not merely complex, but its complexity is coherently oriented towards its *purpose* and object – to enable a human being to know the time. It is a timepiece. It is not just a complicated jumble. It is by virtue of this purpose, which unites and governs its complexity, a purposeful *whole*. This even further removes it from the sphere of chance.

It may be objected that Paley, no less than Darwin, is reasoning from analogy. The difference, however, is that whereas Darwin's use of analogy that between natural selection and artificial, is (as has been pointed out earlier) far-fetched and patently faulty when subjected to even the most superficial scrutiny, that of Paley is so close and cogent that it is tantamount to irrefutable logic, and is in its essential nature not analogy at all. It is simply saying that complex things require a maker. Most people would accept that. Logic isn't necessary to arrive at it. Nor even common sense. It is self-evident.

Now we all know the purpose of a watch. We all know why it was made. But what it may be argued is the purpose of a living being – a plant, or an animal, or man? This is not so self-evident.

Well, on the lowest and most obvious level all these beings are clearly made to survive. They eat, and breathe and reproduce. They are able to survive – as individuals, and as kinds. That's not much of a purpose. But that's as far as Darwin and his followers will go. Or *can* go. It's all about survival, and *only about survival*. That's all. The idea of a possible purpose or value beyond that doesn't come into their ideology. And it *is* an ideology – a belief system about the Universe based on assumptions about the nature of things, materialistic assumptions – which are themselves not scientific. Science, therefore, is not scientific. This is rarely grasped by its practitioners, or its adulators. And when grasped it is quickly suppressed. The ideology of science is in fact contrary to the scientific method. It is thus self-annihilating.

If there *is* a purpose inherent in human life, and for that matter in other kinds of life, beyond the level of mere survival, it is not one (or perhaps many) accessible to the purview of science.

The limitation of science as a mode of enquiry is its basis in positivism. Positivism limits the scope of knowledge to what can be perceived by the senses and extensions of the senses like microscopes and telescopes. It refuses to recognise anything metaphysical and anything that cannot be verified by the scientific method – itself positivistic. This means it refuses, and is in fact *unable* by its own terms of reference, to engage in any critique of the presuppositions upon which it is *itself* founded.

Thinkers have noticed this myopia for a long time, but without doubt the most devastating criticism of positivism, and its willing slave the scientific method, is that of the much-neglected philosopher and thinker P.D. Ouspensky. He put the matter extremely clearly in his remarkable work *Tertium Organum*.

Ouspensky, too, uses the analogy of a watch, but

not for the purpose of arguing the case for design as Paley does, but in order to show the inability of positivism to answer or even approach the question of purpose itself. If a positivist, using only positivistic methods of investigation, were to examine a watch, argues Ouspensky, without the knowledge that it was a device for measuring time, he would certainly see that it is an intricate and complicated piece of machinery, but for him to grasp that it is in fact a *timepiece* he would have to bring in information from *outside* the parameters of strictly positivistic methodology.

Still less would the positivist or scientist, limited as he is to positivist ideology and methodology, be able to understand the purpose of a plant or animal. And even less would he be in a position to comprehend the purpose of a human being.

It's about time we stopped talking and simply looked at examples of what is patently design in nature from the plant and the animal world: though in fact everywhere we look and everywhere we turn we see obvious design, from the humblest to the highest being. So here are a number of illustrations of creatures whose design speaks for itself:

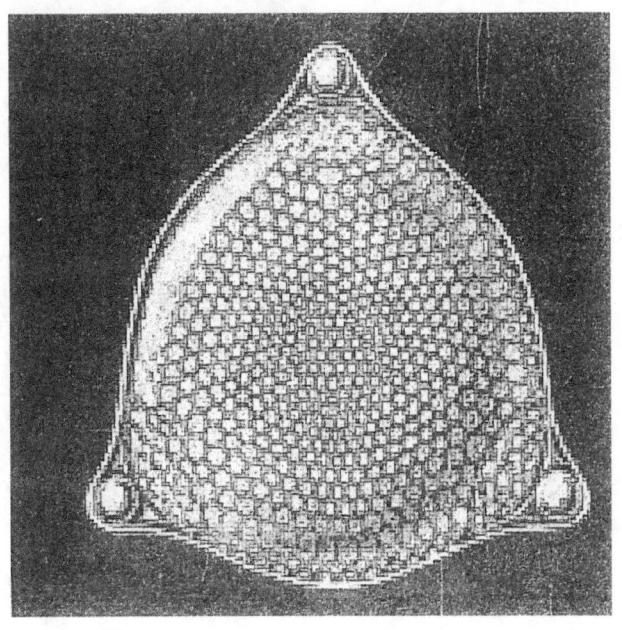

A Diatom.

A minute single-celled plant. These would just cover a full stop. Photograph by Frank Cousins.

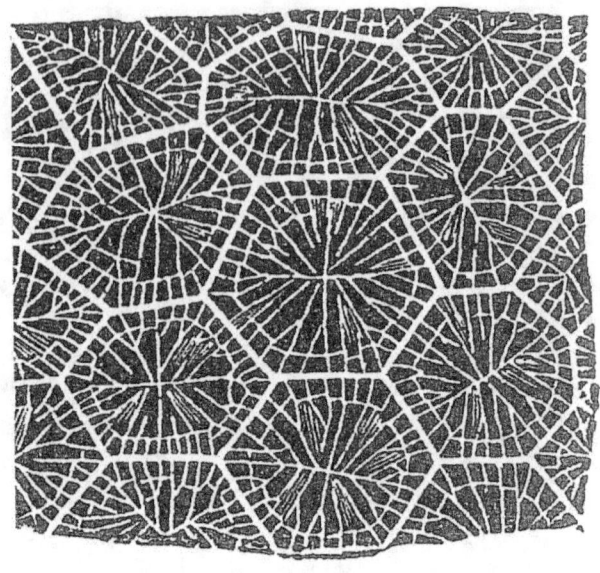

Coral fossils greatly magnified in microphotograph by
Frank Cousins.

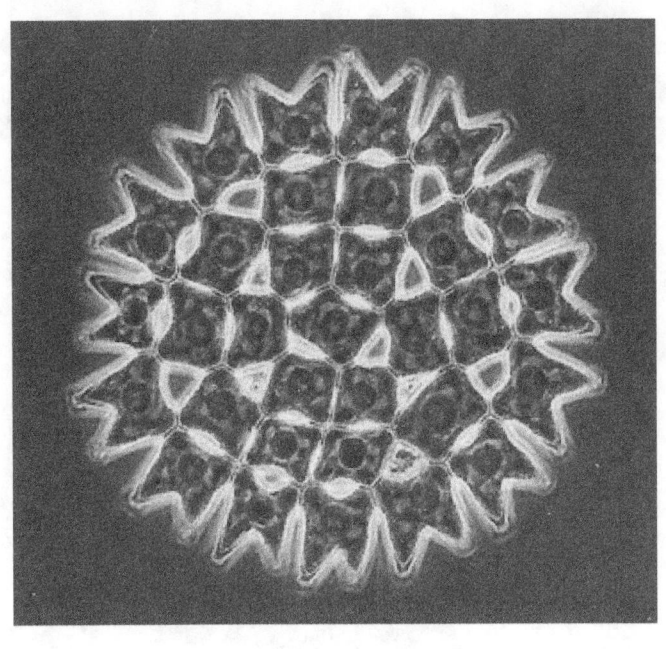

Colony of Algae (single-celled).

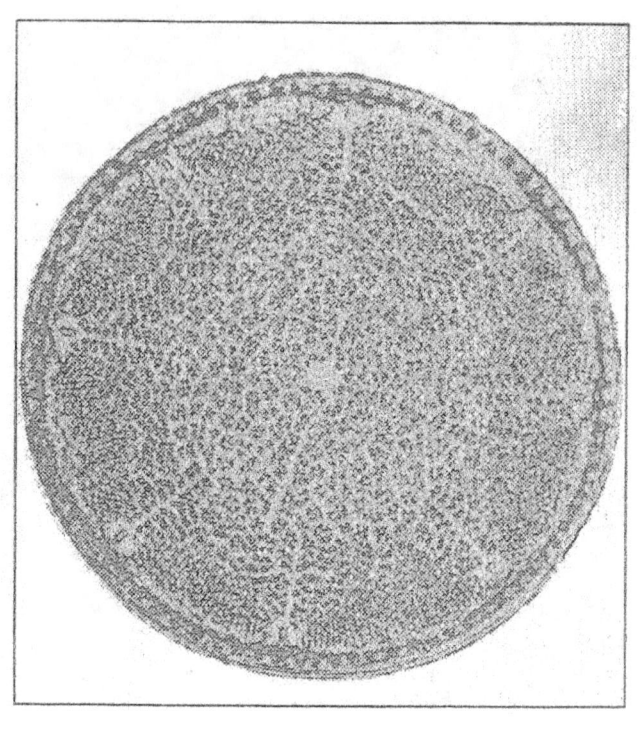

A Diatom.

Microphotograph by Frank Cousins.

Actual size about a full stop.

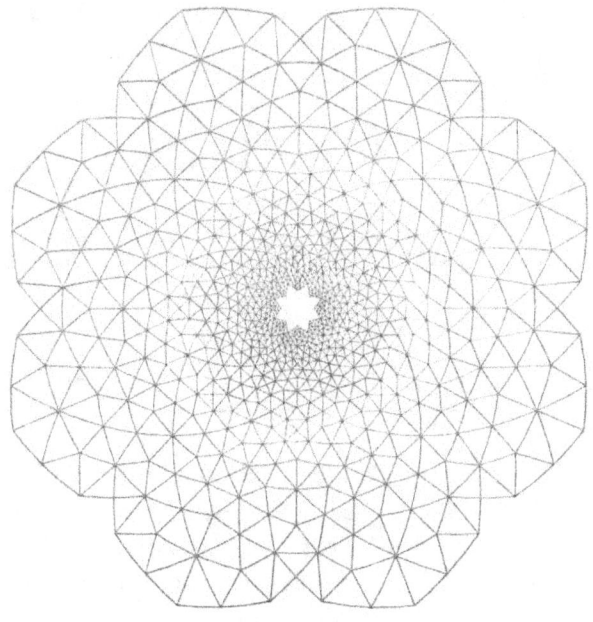

This is an Altair design, named after the star Altair by which mediaeval Arabs navigated. These designs are founded upon and unfold from the same source-pattern as certain Sufi arabesques of the tenth and eleventh centuries which is believed to correspond with an inner template relating to the construction of the Universe.

CHAPTER 8

A PAUSE FOR THOUGHT

I sometimes wonder why I am writing this book at all. It has always been puzzling to me why Darwin's curious theory of evolution should ever have been taken seriously at the time of its inception, in view of the many faults and fissures which it contains which I have pointed out in the preceding pages, and that it has lasted so long. I suspect Darwin himself would have been surprised at its survival. He was never as confident of its "fitness for survival" as his followers, and his uneasiness about the whole proposition is apparent from the beginning.

Many of the reasons for the success of Darwin's theory – if indeed it can be called that, rather than a mere hypothesis and a flimsy one at that – have been discussed, but one which has not been mentioned is the Victorian idea or belief in 'Progress', the dominant mental climate into which his book was launched, and with which his theory of evolution seemed to coincide. 'Evolution' in one form or another was 'in the air', and, for many, Darwin's ideas corroborated their mental proclivities at the time and thus received their support though often without a careful reading of the book itself and a thoughtful consideration of what these ideas really were.

Nevertheless, as has been pointed out, a significant number of his thinking contemporaries were by no means convinced by the Darwinian argument and expressed their disagreement lucidly and powerfully in various statements for the most part forgotten, but which are still as relevant as ever to the debate, with arguments which have never been countered.

And, as has been stressed, their objections to the theory were not based on any faith or belief they might have had regarding the nature of the Universe, but on their knowledge and observation of the natural world, and thinking founded upon them. Most were experts in the same field as Darwin, looking at the same scenario as he was, but disagreeing with his interpretation of what he saw, and identifying serious fault-lines running through his exposition. More often than not he was unable to defend his thesis against the acuity of their criticisms. Now and then he actually admitted (to his credit) to having to change his mind on certain points. For example:

> "Fleeming Jenkin argued in the 'North British Review' against single variations ever being perpetuated, and has convinced me. I was blind and thought that single variations might be preserved much oftener than I now see is possible or probable... I believe I was mainly deceived by single variations offering such simple illustrations, as when man selects."

Letter of Darwin to Wallace

February 2nd, 1869

Darwin was indeed, as he admits, blind. But blind over a much larger field of vision than he identifies in this confession. Nevertheless, he does in this statement begin to recognise the inadequacy of the premise upon whose faulty foundation he proceeds to erect the whole shaky superstructure of his theory – the analogy he draws between domestic selection under man's control, and natural selection under no control at all but the vagaries of chance and environmental circumstance.

It should by now have become obvious to the reader that I am a 'member' of what has been called the 'Intelligent Design' school of thought. In other words, it seems fairly obvious to me that the fantastic intricacy and complexity of the world about us, together with the distinctive individuality of the multitude of the beings that inhabit it, are the work of a Designer, operating perhaps with the help of lesser designers acting as agents.

To believe otherwise, and it is important to recognise that it *is* indeed a belief – as I have endeavoured at length to make clear – seems to be an odd and curiously perverse notion, a wilful flying in the face of facts.

It is not odd, however, if we are aware of the psychological basis – amounting perhaps to an obsession – of this myopia, the paralysis of thought which results from fixation with a particular framework of approach to the world which is not recognised by its adherents for what it is – nothing but a dogma upon which their (the scientists') very self-image depends. Outside that lies an unknown, to believe in the existence of which constitutes a threat to the safety and security of their world-view. To make matters worse, it is an unknown which cannot be known by their ways of knowing. Not a comfortable thought. And one to be rejected or suppressed at all costs. Which includes the cost of knowledge.

Not that this really is, or should be viewed as, a battle for survival. There is nothing wrong with science as far as it goes. All that is being said is that there is a limit to how far it goes. And beyond and above this, there are other ways of knowing.

We use these other modes all the time in our everyday lives when we make value judgements both ethical and aesthetical, and in the normal matters that are most important to us as human beings.

But as has been pointed out, the kind of thinking necessary to elucidate the existence of design in nature and the Universe at large is essentially little more than – and I hate to say it, because it is not the mode of thinking I most admire – logical inference. Design implies a designer.

This is an unavoidable inference. That is surely not asking too much of our mental faculties. The only way to avoid it is to refuse to believe the evidence of our senses, and to fly in the face of the obvious. It is precisely because design implies a Designer that positivist scientists are unable to accept, are in fact 'congenitally' disabled from accepting, the obvious evidence of design all about *us*.

We must now ask the question: Does it matter if there is a designer or not? And the answer is yes, if He has further designs upon *us*.

It is time that we began to address this interesting and significant question.

CHAPTER 9

THE EVOLUTION OF
HUMANKIND

✦━━━━━━━━━━━━━━✦

We must now for the first time look at evolution, *real* evolution, especially the evolution of us, the human race – or more correctly 'human' as far as we know the meaning of that term at the moment – but not necessarily what the full potentialities inherent in it might be.

For Darwin, despite generally accepted notions to the contrary, was not concerned with evolution. He never uses the term 'evolution', and quite deliberately, because he was intent on producing something else and less. He was only interested in change, from so-called simple to more complex forms of life through natural selection. And his central 'value' (if it can be called that) is survival value. That's all.

Now, if evolution means anything meaningful it must mean the evolution of consciousness, of intelligence in the full sense of that word, of awareness of oneself and the Universe one is mysteriously plunged in – the context and medium of one's growth. And growth in any real sense can only be of this kind, the development of what could be called 'being-intelligence'.

This surely must be the essence of evolution, as it must be the essence of whatever we mean by life. It could be called the inner evolution or inner development of being and consciousness.

For several thousand years, perhaps considerably longer, man has had indication of a capacity to increase his consciousness beyond the confines of his ordinary experience, that he has latent potentialities which he has, on the whole, so far neglected, or not recognised for what they were. From time to time, and in various cultures certain individuals have attempted, with not a great deal of success, to inform people of this capacity, and to help them to use it. Some of them have been called spiritual leaders, others have been regarded as teachers and sages, and others have not been recognised at all by the generality for what they were doing.

This is not strange, because what they were, and are, doing is not an easily recognisable activity. If human beings are satisfied with life as they find it, with their consciousness as they find it, they will not even begin to look outside these confines, nor will they even begin to suspect that they *are* confined. Life is alright as it is, isn't it? It's just a matter of making the most of it and looking for what it has to offer. This is merely a matter of living 'fully' within the dimensions one is born into and inhabits – but not conceiving that there might be a greater fullness of life possible in other dimensions.

Nevertheless, more evolved beings have over a very long period tried to tell people of these higher dimensions and the higher life possible in these spheres, which are connected to the ordinary one in a continuum – with success limited by the receptivity of the audience, or as one of their number has put it, whether you have 'ears to hear'.

The central message of these people appears to be that man is in some important sense 'asleep'. What exactly is meant by this term is not necessarily easy to grasp or to communicate. This is the trouble. And as the majority of human beings think that sleep ends when you get out of bed in the morning, they are hardly likely to be interested in the, for them, curious proposition that it does not. And to be told that waking up requires effort, skill and a certain kind of education, is of course likely to fall upon somewhat deaf ears.

People say they are 'not interested'. That is not quite true. Strictly speaking, as they have not understood the situation that is being referred to, they are unfortunately not in a position to claim whether they *are* interested or not. In other words, they are not really qualified to comment upon themselves in this respect.

These among others are the kind of problems which face those attempting to communicate with the sleeping mass of mankind. We can call them the teachers of consciousness, or awakeners of humanity. Theirs is no easy task. And their success in this enterprise has always been limited.

The essence of their teaching is that man has the innate possibility of a higher kind of consciousness than the state he is usually in, and which he accepts as 'normal'. He has this potentiality, but to develop it is not a natural thing. It is in fact to a considerable extent working *against nature*. Nature doesn't want it or need it, being content that man remains under the spell of sleep in which he is cast. Nevertheless, from another point of view it is imperative that as many men and women as possible do learn to wake up, to play their full and real role in the Universe. We can call this the cosmic point of view. From this perspective the evolution of humanity, by which is meant its inner evolution, is a necessity.

From time to time certain men and women appear to have developed their consciousness beyond the 'normal' level which the rest of humanity has taken for granted as 'life'. They have often been called mystics. And are of all times and places – of the East and of the West. Of all religions, or none. For in this enterprise it is not doctrine that matters, but development.

And this development must be deliberate and conscious. Chance plays no role in it. Unlike the notion of "evolution" earlier discussed and criticised.

And, as has also been said, help too is necessary. Help from above, from the more evolved realms of life, of consciousness, of intelligence. Help from higher up the ladder. There must be a reaching down and a reaching up. In many ways the Chain of Being referred to earlier corresponds with this image of the Universe. An image man has temporarily lost or forgotten, and must now find again.

These helpers have had many names. A number of them have been called Sufis. But according to the best authorities Sufism is not to be regarded as 'Islamic mysticism' – as it is commonly misdescribed and misunderstood. It is in fact a free-moving and versatile teaching of which that is but a local branch – not the tree itself. It has existed both before and beyond the confines of Islam, not bound by culture or religion, neither of the East nor the West, but addressing itself to the essential human being.

The misconception referred to has been well put by Robert Ornstein in his valuable work *The Mindfield*:

> 'The belief that Sufism must be Islamic is a mere Middle Eastern ethnocentrism, the confusion of yeast itself with one particular loaf of bread.'

A leading contemporary Sufi has said:

> 'The Sufis affirm that the organism known generally as Sufism has been the one stream of direct evolutionary experience which has been the determining factor in all the great schools of mysticism.'

Considerable research tends to bear this out.

Note well that it is *experience* which is stressed. This is because the essence of this activity is spiritual rather than religious, concerned with experience rather than belief. For while belief binds, experience liberates. It is relevant that the word religion is derived from the Latin 'religere' – to bind. Sufism therefore is not a religion. It has concepts rather than beliefs, and concepts that must be *realised* in experience, in consciousness. It is concerned with development – the development of human understanding.

The Sufi master Ustad Hilmi makes the following statement:

> 'Man, we say we know, originates from far away; so far indeed, that in speaking of his origin, such phrases as 'beyond the stars' are frequently employed. Man is estranged from his origins. Some of his feelings (but not all of them) are indicators of this.
>
> Man has the opportunity of returning to his origin. He has forgotten this. He is, in fact, 'asleep' to the reality.
>
> Sufism is designed as the means to help man awaken to the realisation, not just the opinion of the above statements. Those who awaken are able to return, to start the 'journey', while living this present life in all its fullness. Traditions about monasticism and isolation are reflections about short-term processes of training or development, monstrously misunderstood and grotesquely elaborated to provide refuges for those who want to stay asleep... Since Sufism depends upon effectiveness, not belief.
>
> If man finds himself again, he will be able to increase his existence infinitely. If he does not, he may dwindle to vanishing point.
>
> People have been sent from time to time, to try to serve man and save him from the 'blindness' and 'sleep', which today would be better described as 'amnesia' which is always described in our technical literature as a local disease. These people are always in touch with the Origin. They have been of all races, and have belonged to all faiths, or none.'

They offer what might be termed 'evolutionary education'. But teaching depends on learning, and a willingness to learn how to learn. Not everyone is willing to adopt the necessary learning posture, to make themselves available to the teaching. Which requires a certain humility, a certain patience – which to the Sufis are learning techniques rather than virtues.

And as can be seen in the words of Ustad Hilmi, the Sufi has to remain in society to follow his path. He must 'be in the world but not of the world'. That is how he serves himself and serves other people. Another master has explained why. 'For the Sufi the world is a fashioning instrument.' It is in every sense, and literally 'a growing medium'.

As Akbar Khan puts it in his Tasawuf-i-Azim:

> 'Man is destined to live a social life. His part is to be with other human beings. In serving Sufism he is serving himself, serving society, and serving the Infinite. He cannot cut himself off from any of these obligations and become or remain a Sufi. The only discipline worthwhile is that which is achieved in the midst of temptation. For power is that which is won through being wrested from the midst of weakness and uncertainty.'

Idries Shah, a Sufi master, goes further:

> 'To the Sufi the evolution of the Sufi is within himself and also in relationship with society. The development of the community, and in fact the destiny of all creation - even nominally inanimate creation - is interwoven with the destiny of the Sufi.'

And Rumi reminds us:

> '...the new faculties for which man yearns, generally unknowingly, will only come into being if he now takes part in the development of his own evolution.'

Arkon Daraul, in his book *Secret Societies* makes the following observations about a Sufi school he has encountered:

> 'The People of the Path believe that a certain nobility of mind and purpose resides within every human being. This is the task of the Sufi teacher to discover and develop in the individual... Members of the Order believe that it gives them something which they have unconsciously sought for years.
>
> Mankind, they say, has certain ideas, certain capabilities, certain capacities for experience. These things are all related. The goal is the Ideal Man, who shall use every aspect of his experience to be "In the World and yet not of the world".'

It is clear that these various statements about the nature and objectives of the Sufi mystics concur.

Probably the best definition of a mystic – bearing in mind Oscar Wilde's brilliant warning that 'to define is to limit' – is Evelyn Underhill's: 'Mysticism is the art of union with Reality. The mystic is the person who has attained that union in a greater or lesser degree.'

There we have it.

She refers to a considerable number and variety of mystical experiences in her extraordinary book *The Mystic Way*, as does F.C. Happold in his valuable and wide-ranging study and anthology *Mysticism*. What is interesting and significant is the extent to which the mystical states described by both researchers, whether the experiencers are of the East or the West, are similar in their essential characteristics. The striking sameness and consistency of what the mystics tell us indicates it is not imagination. As Mr. Happold observes:

> "Not only have mystics been found in all ages, in all parts of the world, and in all religious systems, but also mysticism has manifested itself in similar or identical forms wherever the mystical consciousness has been present."

And, he concludes:

> "May we not see in the mystics the forerunners of a type of consciousness which will become more and more common as mankind ascends higher and higher up the ladder of evolution?"

One of the people Happold includes in his community of mystics is the significant thinker and priest Pierre Teilhard de Chardin, author of *The Phenomenon of Man*, and *Le Milieu Divin*. Significant because his knowledge also extended to the unpriestly fields of geology and palaeontology and biology, his studies of which he reconciles with his own particular image of Christianity in a fusion which expresses a vivid and compelling view of evolution. Here we have the most interesting and unique phenomenon of a man of the spirit who is also a distinguished scientist and is able to unite his knowledge in both areas to provide a vision of the Universe that connects matter and spirit in one organic whole.

This is the kind of thinking we need. Or, in particular, the West needs. It breaks the deadlock between religious dogma on the one hand, and scientific dogma on the other. By rising above both. De Chardin is not the slave of either ideology, nor paralysed by either paradigm. So we can see why, while orthodox scientists could not accept his mystical speculation, the Church (he was a Jesuit) withheld permission to publish those of his books which combined his spiritual and scientific ideas, which is the distinctive essence of his work. De Chardin's purely scientific papers, of which there were many, did not suffer from the non imprimatur.

He sees the Universe as an evolutionary process, always moving towards greater complexity, but this has simply been the *vehicle* for a movement towards higher levels of consciousness. This, for De Chardin, is the point of it all. Consciousness is everything. The be all and end all of existence. The Universe bears witness to a design, and that design is developmentally oriented. It is directed by God – but man has a role in this too. He would agree with the Sufis – man must participate in his own evolution. He must learn to help himself.

Happold himself comes to the following conclusions:

> "The evolution of life on this planet may be envisaged as a process of organic involution upon itself, as a movement from the extremely simple to the extremely complex. In this involution of complexity there has been a steady increase in interiorization, an enlargement of the consciousness. As they become more complex, structures become more vitalized and more conscious.
>
> It is difficult to avoid the conviction that there must be some pattern of organization inherent in the whole evolutionary process, a dynamic inclination, whatever its origin, impelling structures to behave in certain ways and to develop in certain directions.
>
> I think that Father de Chardin may, as a biologist and a palaeographer, have glimpsed one of those

> superb syntheses which have marked the
> development of human thought. His general picture
> of the shape of evolution, as a process of expanding
> interiorization, seems to ring true."

Anyone who reads his book will see that Happold arrives at this conception of the Universe as a result of long and deep study and reflection, and it is interesting that he regards the mystics as the forerunners of the next stage in human consciousness. There is however nothing 'extremely simple' about any kind of life on this planet, as I have stressed earlier, and this is the only point on which I would have to disagree with him.

Nevertheless, one point must be stressed. However elevated and illuminating are the experiences which mystics report, it seems that on the whole they are temporary, sporadic or sometimes accidental. They are experiences. Whereas it appears that the Sufis and certain other schools of thought maintain the real possibility of a permanent state of consciousness in man – a change in his very being, that will sustain a different and superior awareness of both himself and the world around him. This is the real aim, they aver. It therefore goes beyond mysticism as usually understood, for the accent is not on special experience or revelation, wonderful though these may be, but on the deliberate cultivation and attainment, through education and effort, of a higher kind of life, a higher kind of man.

Though many people do not believe in the existence and activity of a higher kind of human being on the planet, unknown on the whole to the rest of us, there are nevertheless hints and indications throughout history of such a presence. In all the great cultures it is possible to discern evidence of another humanity hidden within the humanity we think we know.

G. I. Gurdjieff, himself a remarkable man, who made a profound and indelible impression on all who met him, claimed to have made definite contact with beings of this order, from whom he learnt much:

> "The humanity to which we belong, namely the whole of historic humanity known to civilization, in reality consists of only the outer circle of humanity within which there is another inner circle."

Responsibility for the evolution of humanity, he explained, was vested in this group. By evolution was meant "the evolution of human consciousness". He added:

> "If the evolution of humanity can only proceed through the evolution of a certain group, which in its turn, will influence and lead the rest of humanity."

One does not have to believe Gurdjieff if one doesn't want to. But for some of us therein lies the hope that we can transcend the defects and inadequacies of the present human condition.

Ouspensky put it like this:

> "If the existence of hidden knowledge is admitted, it is admitted as belonging to certain people, but to people we do not know, to an inner circle of humanity."

"According to this idea, humanity is regarded as two concentric circles. All humanity which we know and to which we belong form the outer circle. All the history of humanity that we know is the history of the outer circle. But within this circle is another, of which men of the outer circle know nothing, and the existence of which they only sometimes dimly suspect – although the life of the outer circle in its most important manifestations, and particularly in its evolution is actually guided by the inner circle. The inner or the esoteric circle forms, as it were, a life within life, a mystery, a secret in the life of humanity."

Another way of looking at this is presented in the Sufic fable given to us by Idries Shah as the opening chapter of his book *The Sufis*. It is called 'The Islanders'. He introduces is it as follows:

'Most fables contain at least some truth, and they often enable people to absorb ideas which the ordinary patterns of their thinking would prevent them digesting. Fables have therefore been used, not least by the Sufi teachers, to present a picture of life more in harmony with their feelings than is possible by means of intellectual exercises.

Here is a Sufic fable about the human situation, summarised and adapted, as must always be, suitably to the time in which it is presented.

Once upon a time there lived an ideal community in a far off land. Instead of uncertainty and vacillation, they had purposefulness and a fuller means of expressing themselves. Although there were none of the stresses and tensions mankind now considers essential to its progress, their lives were richer because other, better elements replaced these things.

Theirs, therefore, was a slightly different mode of existence. We could almost say that our present perceptions are a crude, makeshift version of the real ones which this community possessed.

They had real lives, not semi-lives.

We can call them the El Ar people.

They had a leader, who discovered that their country was to become uninhabitable for a period of, shall we say, twenty thousand years. He planned their escape, realizing that their descendants would be able to return home successfully, only after many trials.

He found for them a place of refuge, an island whose features were only roughly similar to those of the original homeland. Because of the difference in climate and situation, the immigrants had to undergo a transformation. This made them more physically and mentally adapted to the new circumstances; coarse perceptions, for instance, were substituted for finer ones, as when the hand of the manual labourer becomes toughened in response to the needs of his calling.

In order to reduce the pain which a comparison between the old and new states would bring, they were made to forget the past almost entirely. Only the most shadowy recollection of it remained, yet it was sufficient to be awakened when the time came.

The system was very complicated, but well arranged. The organs which were really constructive in the old homeland were placed in a special form of abeyance, and linked with a shadowy memory, in preparation for its eventual activation.

Slowly and painfully the immigrants settled down, adjusting themselves to the local conditions. The resources of the island were such that, coupled with effort and a certain form of guidance, people would be able to escape to a further island, on the way back to their original home. This was the first of a succession of islands upon which gradual acclimatization took place.

The responsibility for this "evolution was vested in those individuals who could sustain it. These were necessarily only a few, because for the mass of the people the effort of keeping both sets of knowledge in their consciousness was virtually impossible. One of them seemed to conflict with the other. Certain specialists guarded the "special knowledge". This "secret", the method of effecting the transition, was nothing more or less than the knowledge of maritime skills and their application. The escape needed an instructor, raw materials, people, effort and understanding. Given these, people could learn to swim, and also to build ships.

For a time things went well, and the people in charge of the escape operation found there were enough islanders interested to seek their assistance in enabling them to make the voyage from their place of exile.

However a time came when a man who had been found, for the time being, lacking in the necessary qualities rebelled against this order and managed to develop a masterly idea. He had observed that the effort to escape placed a heavy and seemingly unwelcome burden upon the people. At the same time they were disposed to believe things which they were told about the escape operation. He realized that he could acquire power, and also revenge himself upon those who had undervalued him, as he thought, by a simple exploitation of these two sets of facts.

He would merely offer to take away the burden, by

affirming that there was no burden.

He made this announcement:

"There is no need for man to integrate his mind and train it in the way which has been described to you. The human mind is already a stable and continuous, consistent thing. You have been told that you have to become a craftsman in order to build a ship. I say, not only do you not have to become a craftsman - you do not need a ship at all! An islander needs only to observe a few simple rules to survive and remain integrated into society. By the exercise of common sense, born into everyone, he can attain anything upon this island, our home, the common property and heritage of us all!"

The tonguester, having gained a great deal of interest among the people, now "proved" his message by saying:

"If there is any reality in ships and swimming - show us ships which have made the journey, and swimmers who have come back!"

This was a challenge to the instructors which they could not meet. It was based upon an assumption of which the bemused herd could not now see the fallacy. You see, ships never returned from the other land. Swimmers, when they did come back, had undergone a fresh adaptation which made them invisible to the crowd.

Nevertheless the escapers did their best to argue with the revolt and to try to communicate both the importance and the feasibility of their operation to the people.

But with very limited success. The new gospel was welcomed on all sides as one of liberation. Man had discovered that he was already mature! He felt, for the time being at least, as if he had been released from responsibility.

Society had now temporarily equilibrated itself within the island, and seemed to provide a plausible completeness – if viewed by means of itself.

Since the skills of shipbuilding had no obvious application within the society, the effort could easily be considered absurd. Boats were not needed – there was nowhere to go. The consequences of certain assumptions can be made to "prove" those assumptions.

An entry in the great Island Universal Encyclopaedia show us how the process worked:

SHIP: An imaginary vehicle in which impostors and deceivers have claimed it possible to "cross the water", now scientifically established as an absurdity. No materials impermeable to water are known on the Island, from which such a "ship" might be constructed, quite apart from the question of there being a destination beyond the Island. Preaching "shipbuilding" is a major crime under Law XV11 of the Penal Code, subsection v1 The Protection of the Credulous. SHIPBUILDING MANIA is an extreme form of mental escapism, a symptom of maladjustment. All citizens are under a constitutional obligation to notify the authorities if they suspect the existence of this tragic condition in any individual.

See: Smith, J., Why "Ships" Cannot be Built, Island University Monograph No. 1151.

Attitudes like this obviously made it both difficult and dangerous for the instructors to do their work and therefore they were forced to continue their activity more and more secretly. They had to adopt various disguises or 'go underground'.

However, conditions on the island did not entirely fill these dedicated people with dismay. After all, they had originated in the very same community, and had indissoluble bonds with it, and with its destiny.

Clandestinely the ships still raised their sails, the swimmers continued to teach swimming....

The fable has not ended, because there are still people on the island.

But if you told most of them this tale, they would smile and say, "Well, it's only a fable, isn't it?"'

The Contribution of G. I. Gurdjieff

At this stage it is interesting to consider the contribution of that remarkable man George Ivanovitch Gurdjieff to certain other matters relevant to the discussion.

Without doubt, Gurdjieff possessed unusual knowledge about the role of man in the Cosmos he finds himself in, which he said he acquired from a Sufi monastery somewhere in the Hindu Kush mountains about the beginning of the 20th century. The record of his unrelenting search for truth can be seen in his autobiographical book *Meetings with Remarkable Men*.

As a youth, Gurdjieff became obsessed with the idea that there was a purpose and aim behind human life which was hardly ever glimpsed in the ceaseless generations of man. He was convinced that in former epochs certain men had possessed genuine knowledge of such matters, and that this knowledge was still preserved, somehow, somewhere.

Together with a number of others like-minded

with him, Gurdjieff began a search (lasting decades) for traces of this knowledge. His "society" of seekers, sometimes singly sometimes in small groups undertook journeys to remote places where traces of this ancient knowledge might survive. Their survey took in Persia, Turkestan, Egypt, Tibet, India and Afghanistan. Some disappeared, some did not survive the hazards of these often difficult explorations.

However, it seems clear from his own account that Gurdjieff and one of his fellow seekers did, in the end, find what they were looking for in a Centre somewhere in the remote recesses of the Hindu Kush mountains. There seems little doubt that this is the very same monastery visited later in 1965 by Desmond R. Martin who reported his impressions in an article in the *Times* that year. All visitors appear convinced of the authenticity of kind of knowledge and information they were given access to.

Later, during the period 1914 to 1917, Gurdjieff conveyed some of this knowledge to his pupils in Moscow. We have mentioned Gurdjieff's statements regarding the evolution of a higher kind of human being. We will now look at information he acquired of a cosmological and planetary nature which is of considerable interest and relevance to the present subject.

First, there is Gurdjieff's statement about the role of organic life on Earth. To understand it fully one has to realise that a central theme of his teaching was that the Universe with all its galaxies, solar systems, stars, planets and moons was a vast connected, organically evolving whole. All were bound together and influence each other, by a fantastic network of interflowing vibrations in a spectrum of vibrations from coarse to fine, and ever finer. The Universe was one pulsating Whole.

This then is what Gurdjieff said:

> "The conditions to ensure the passage of forces are created by the arrangement of a special contrivance between the planets and the Earth. This contrivance, this 'transmitting station of forces', is organic life on Earth. Organic life was created to fill the interval between the planets and the Earth.

> "Organic life represents, so to speak, Earth's organ of perception. Organic life forms something like a sensitive film which covers the whole of the Earth's globe and takes in those influences coming from the planetary sphere which otherwise would not be able to reach the Earth, the vegetable, animal, and human kingdoms are equally important for the Earth in this respect. A field merely covered with grass takes in planetary influences of a definite kind and transmits them to the Earth. The same field with a crowd of people on it will take in and transmit other influences. The population of Europe takes in one kind of planetary influences and transmits them to the Earth. The population of Africa takes in planetary influences of another kind, and so on.

> "All great events in the life of the human masses are caused by planetary influences. They are the result of the taking in of planetary influences. Human society, unknown to it, is a highly sensitive mass for the reception of planetary influences. And any accidental small tension in planetary spheres can be reflected for years in an increased animation in one or another sphere of human activity. Something accidental and very transient takes place in planetary space. This is immediately received by the human masses and

affects their action and behaviour in one way or another to an extent which is not realised or even guessed at by the inhabitants of the planet. Nor would it be believed if it were explained to them.

"Organic life is the organ of perception of the Earth and it is at the same time an organ of radiation. With the help of organic life, each portion of the Earth's surface occupying a given area sends, every moment, certain kinds of rays in the direction of the sun, the planets, and the moon. In connection with this, the sun needs one kind of radiation, the planets another kind, and the moon another. Everything that happens on the Earth creates radiations of this kind. And many things often happen just because certain kinds of radiation are required from a certain place on the Earth's surface."

Not everyone will believe, or want to believe, this kind of information – as Gurdjieff was always fully aware. Yet for me at any rate it seems to ring true, it makes a strange kind of sense. There is a convincing authenticity about it. I have always wondered why certain cities have grown up exactly where they have on the surface of our planet, the great ones being in effect concentration points for vast numbers of human beings, and I have never been convinced that merely chance, or only geography were sufficient explanations for their presence and position, and so this explanation is both interesting and credible to me. I can quite imagine Rome, or London, or Paris or New York as great power-houses, huge transmitting and receiving organs in some great cosmic web and interplay of vibrations, nodes in a network unknown to us.

From this perspective, too, one may regard the large-scale destruction and depletion of the great rainforests that we are now witnessing, not only as the irreparable loss of the irreplaceable beings whose unique habitat they are, but also as the obliteration of what must surely be significant receiving and transmitting organs in the sensitive surface film of the planet playing the role which Gurdjieff describes. He also warned that if the balance is disturbed by serious perturbations of such a kind, Nature was prone to redress the balance in ways that were not predictable by us.

It can be seen that by using the term Earth's 'organ of perception' to describe organic life, Gurdjieff is implying that the planet is a living being. According to the teaching he was expounding, this is indeed so. Each planet is, he averred, "an animate, intelligent being" – with a definite lifespan.

This idea may be strange for us but it was not at all so to many earlier civilizations. The Greeks in particular perceived the wandering orbs of the solar system in this light, as we all know.

More recently, the concept of Gaia, created and developed by James Lovelock in the 1960s, and argued in his book in 1979 *Gaia: A new look at life on Earth* envisages the planet as a living being regulating the balance, composition and complex interplay between the biosphere, hydrosphere and atmosphere, to constantly provide an environment or medium suitable for life. He chose Gaia as this was the Greek primordial goddess of the Earth.

Lovelock came to this conclusion as a result of his work as an independent scientist when he observed certain curious facts about the planet that were not explainable in any other way.

Firstly, the global surface temperature of the Earth has remained constant, despite an increase in the energy provided by the Sun.

Secondly, the composition of the atmosphere remains constant, even though it should be unstable.

Thirdly, the salinity of the ocean is constant, despite the continual influx of river salts.

Fourthly, without the constancy of these three elements, the planet would not be habitable for life as we know it.

Since life started on the Earth the energy provided by the Sun has increased by 25% to 30%. However, the surface temperature of the planet has remained remarkably constant when measured on a global scale. Why and how is this the case?

The atmospheric composition of the Earth is currently constant, consisting of 79% nitrogen, 20.7% oxygen and 0.03% carbon dioxide. These proportions do not vary. Why not?

Ocean salinity has been constant at about 3.4% for a very long time, and this has been a long-standing mystery because river salts should have raised it much higher than observed. Now, salinity stability is very important as most cells require a rather constant salinity and do not generally tolerate more than 5%. Without such stability most life in the sea would be at risk.

I could go on. You'd better read Lovelock's book if you want to know more about this interesting planetary phenomenon. The Earth certainly appears to be in some way governing important aspects of its surface environment to enable it to support life.

Lovelock describes this as the maintenance of a 'homeostasis' for making conditions on the planet 'hospitable' for the presence of living beings upon it.

But this state of affairs is itself produced by these very living beings. In other words, they create and maintain the very environment they require to survive in. This is more than 'the balance of nature' as normally understood, which is really a 'mechanical' concept. It is rather a concept of an organic, dynamic, holistic, interacting system, implying direction from a source above and beyond the material with which it deals.

His book *Gaia: A New Look at Life on Earth* is replete with examples of these self-regulating processes, and here I will merely mention one that has always interested me – the role of the ocean algae in the history of life on the planet. Firstly it was thanks to their photosynthesising ability that in Precambrian times the atmosphere was given the oxygen necessary to support the later development of life on Earth. And secondly it is largely due to their role in absorbing and removing carbon dioxide from the atmosphere that its proportion of the composition of the air has been, and on the whole still is, constant at about 0.03 per cent. Or 35/1,000 of 1 per cent of the atmosphere. We are now aware how serious the consequences can be of even a very slight increase in that content in the form of the global warming we are now witnessing. Significantly however, considerable increases in the ocean algae have been detected. Needless to say, this does not mean that man need do nothing about the problem of his own making. Of course it could be argued that

Gaia also works through the human population.

There is also the nitrogen cycle which has been observed for a long time, but has not been appreciated and recognised for what it is – another aspect of the maintenance of both the content of the atmosphere and its relationship with the biosphere with particular regard to the flow of nitrates through it, so necessary as nutrients for plants and thus for all animals. We have already noted the vast intake of nitrogen directly from the atmosphere by the nitrogen-fixing bacteria by blue-green alga in the sea and mainly the legume family on land. What we have not mentioned is the concomitant release of this nitrogen by the activity of the nitrogen-releasing bacteria at the other end of the cycle returning it to the atmosphere to enable the proportion of this gas in its composition to remain constant at 79%.

The point is this. How can such co-ordination throughout the biosphere be the product of chance? There has to be a 'meta-regulator' at work to account for this overall state of affairs. Some controlling agent is operating.

It certainly supports Gurdjieff's assertion that in some sense our planet is a living being.

And reflecting further on the concept of the role of the film of organic life on the Earth's surface as its organ of perception, with its mosaic of different areas and their inhabitants all contributing their special role in the great vibratory web, whether fish, field or forest, could not that be the reason almost the whole planet is populated with animals and plants in every niche and in every environment, perfectly adapted to the most varied and often inhospitable areas? Would one *expect* to see penguins in the frozen places of Antarctica, bats living in caves, algae in hot springs, mangrove forests in salt water, or for that matter all the fish of the sea with their extraordinary ability to breathe under water – if they were not *there*? One could go on almost indefinitely.

CHAPTER 10

CONCLUSION

The time has come to sum up.

I have advanced a number of reasons why the Darwinian hypothesis cannot be accepted as satisfactory despite its curious hold on our culture for so long. This is not true of other cultures – particularly in the East. But even here in the West, apart from the view of hard-line scientists, I have suggested that it is the general concept of evolution that has been accepted rather than the particular method – 'natural selection' – by which Darwin and his adherents believe it to have taken place.

Most people haven't read his book and have not given the idea much thought. Nor are they aware how pivotal and crucial it is to his whole theory. When they do, they tend to be surprised at the simplistic and crude naivety of this 'concept' as an explanation for the shaping of species in all their wonderful variety, both plants and animals – still less, human beings. When they do, they are surprised to realise how far-fetched it all is.

Very often people tend to confuse the undeniable fact of a hierarchy in the scale of living beings, central to the concept of the Great Chain of Being, with the proposition of gradual evolutionary development. The Chain of Being doesn't necessarily imply the *change* of beings.

Let us now identify the main points that have been made in this critique of Darwinism.

Firstly, everything we see and encounter about us appears to be whole and perfect in its kind, whether bird of the air, fish of the sea, tiger of the forest, snake of the rock, ant on the ground, and bacteria in the soil. And the more we learn about them the more we realise how wonderfully they are wrought. This surely stares us in the face – unless one has deliberately turned one's face the other way. Every being declares its design, and that design is unique. It is a far greater act of 'faith' – if we can call it that – to believe that this is all the product of chance than to believe we are witnessing the works of conscious creation. Design implies a designer.

We see too how beings are not, cannot be, isolated, but all subsist in networks of interrelationships and interdependence, a web of life, most obvious in small-scale symbiosis examples of which we have pointed to, less obvious in the larger ecosystems which support clusters of individual creatures, and finally in the wider ecosystem known as the biosphere in which bacteria, plants, animals and we humans are all subsumed for our mutual survival.

Wherever we look, we see pattern, and where we

see pattern it is a fair assumption to posit the existence of purpose.

Thirdly, there is the conspicuous lack of intermediates between species so necessary to the credibility of Darwinian theory which was pointed out by able and knowledgeable critics at the time of the publication of *The Origin of Species*, continues to be pointed out today despite all the geological investigation since, and which is grimly admitted even by its supporters. It is indeed a thorn in their flesh. The hoped-for sequences of intermediary creatures refuse to materialise.

It has become patently clear that the 'gaps' in the geological record are not gaps in the record – but the geological record is a record of gaps. That, like it or not, *is* the record.

To put it another way, the lack of intermediaries are gaps in the theory rather than gaps in the fossil record. For the record is *one* of gaps – though complete in itself. This lack constitutes not just a lack of the evidence necessary for the validity of the theory. It actually contradicts it. And undermines it. Darwin's worst fears have been realised. The state of the geological record which always made him uncomfortable. It refused to blur the distinctions between species in the way his theory demanded. But he had always hoped that future investigations would provide the evidence he was looking for. Yet 150 years of digging has not come up with even a few of the vast range of intermediates needed to indicate the evolutionary continuity he *had* to believe in.

Of course, all this reflects and reinforces the striking *distinctness* of the planet's creatures by the individual separateness and isolation that the record reveals.

Connected with this, and corroborating it, is the remarkable evidence of the continued unchangedness, the persistence, of creatures to remain as they are throughout eras of geological time. We have noted how the humble alga of the sea has retained its form and nature for many millions of years since it first began playing its indispensable role in the ocean's ecosystem, and providing oxygen for the terrestrial animals of the planet and ourselves since Cambrian times. It is the same today as it was then and always has been. It has sustained its selfhood and refused to alter one bit, despite all the vicissitudes of its environment it has had to face over such vast periods. So has the starfish. And the sea urchin. And countless other inhabitants of the sea and land. The case of the coelacanth fish simply highlights an innate propensity common to all creatures – the indelible declaration of their unique and constant identity.

The ant in the amber, the bat in the cave, the fern by the drainpipe, the lichen on the rooftop, all proclaim their intrinsic self-persistence and self-sustaining ability through the vagaries and vicissitudes of many millions of years. That wonderful Magnolia tree you pass every morning has been exactly the same for twenty million years.

This innate stability of living things, this adherence to their own identity, contradicts the tendency to chance variation combined with survival selection to produce any appreciable difference beyond a certain point close to the original. It is, however, the central tenet of Darwinian ideology that there is no end to the potential of chance variation – nowadays called mutation – to alter beings to such an extent that they turn into a different species! But not only is all evidence to the contrary ignored, but also the well-known phenomenon of reversion to type – which is of course an expression of the intrinsic stability being referred to.

We have shown earlier how one of Darwin's most acute critics, and one who considerably disturbed him, Professor Henry Jenkin, having studied the limits of variation possible through domestic selection and applying his findings to wild animals and plants concluded that there was what he termed 'a sphere of variation' within which the range of possible changes of this kind were contained. This concept complements and corroborates the point made above – that there is an inherent inbuilt boundary to the amount of variation a creature can undergo whether left to itself or even whether not left to itself.

As Jenkin says:

> 'These differences are extremely small as compared with the range required by his theory, but he assumes that by accumulation of successive differences any degree of variation may be produced. He says little in proof of the possibility of such an accumulation, seeming rather to take it for granted... That theory rests on the assumption that natural selection can do slowly what man's selection does quickly; it is by showing how much man can do that Darwin hopes to prove how much can be done without him.'

But, Jenkin continues, referring to the wild pigeon which he very generously and rather dubiously allows could be the possible 'parent' of the various fancy pigeons as Darwin believed:

'When we have granted that the 'struggle for life' might produce the pouter or the fantail, or any divergence man can produce - we need not feel one whit more disposed to grant that it can produce divergences beyond man's power. The difference between six years and six myriads, blinding by a confused sense of immensity, leads men to say hastily that if six or sixty years can make a pouter out of a common pigeon, six myriads may change a pigeon into something like a thrush.'

Common sense supports the professor's proposition of a 'sphere of variation'.

We have also been constrained to point out two major defects in Darwin's thinking which undermine his thesis as a whole — and from beginning to end. The first is his use of an analogy, the validity of which is ably criticised by Jenkins as we have just seen, which while perfectly acceptable and in fact an essential mode of operation in some kinds of thought, cannot be accepted as the foundation proposition, the very basis, for a scientific theory or argument.

The second is the making of an elementary error in logic which one doesn't have to have studied the subject at Oxford not to make. It is of course the fallacy of *post hoc ergo propter hoc* — after this therefore because this — which Darwin and his followers have become willing victims of, in order to interpret the chronological succession of plants and animals in the record of geology to make it accord with the assumptions of their theory. It is however understandable in the context of their ideological conditioning.

These defects of thinking are all the more serious in that they act as major premises upon which the whole theory of natural selection is erected. It is neither scientifically nor logically sound. It is indeed surprising that it has been believed in unquestioningly by so many for so long.

Not long ago I was talking to a farmer in one of his fields where a herd of sheep were grazing, and asked him what he thought about the theory of natural selection having outlined it to him as well as I could. "You must be joking!" he said. "You can make that there sheep into a better sheep, but you can't turn it into a cow, can yer? This fellow, Darwin – he must've bin a bit daft!"

I am inclined to agree with the farmer.

Looking, too, at Darwin's central motive force – for the evolution of species, the struggle for survival – one cannot fail to be struck by its sheer inadequacy to fulfil this purpose. 'Survival' is much too low-level an objective to act as an agent for development beyond an early cut-off point. It lacks the necessary dynamic to carry a creature forward to a superior mode, to a superior stage – as the theory demands. It is essentially static rather than dynamic. Its requirements are stasis and stability. As has been pointed out, if this was the prime objective the bacterium would not have any further incentive but to remain a bacterium. It survives perfectly well as it is, and was all those billions of years ago. The same applies to the alga in the ancient ocean. It was utterly self-sufficient, and able to take care of itself. Neither creature could have produced the hierarchy of animal or plants respectively if survival was the only operative factor.

There was no need for anything else.

No struggle for survival garnished by a sprinkling of chance can account for all the great 'leaps' between bacteria and barnacle, mollusc and fish, fish and amphibian, amphibian and reptile, reptile and mammal. Their differences are so obvious and striking that they cannot be explained away in this way.

And with regard to the kingdom of plants, the extraordinary process of photosynthesis, while it undoubtedly *enables* survival, cannot be *explained by* survival.

Nor can the arrival of the mosses and liverworts (the bryophytes) be explained credibly as the result of natural selection among the algae. Not only are their differences pronounced, but in any case, the algae have shown such an incredible ability to adapt to almost every imaginable terrestrial and aquatic environment on the planet without changing their essential nature, that they would not need a development of this kind to maintain their existence.

Again, the arrival of the vascular plants, with their water-conducting faculty enabling tall plants to grow, particularly trees, cannot be explained as the meeting of a need for survival, or indeed in itself. And as was true of the alga and bryophytes – no transitional forms have been found.

As for the mysterious arrival of the flowering plants in sudden profusion in the Cretaceous period, 140 million years ago, that is virtually a well-known fact which has long troubled the sleep of those disposed to the Darwinian concept of evolution, as indeed it did that of its producer.

We are therefore constrained to repeat the admission of the respected botanist Dr. E.H. Corner of Cambridge University:

> "Evidence can be adduced in favour of the theory of evolution.... but I still think, to the unprejudiced the fossil record of plants is in favour of special creation...The evolutions must be prepared with an answer, but I think that most attempts to answer would break down before an inquisition."

One suspects that Darwin was himself aware of the inability of chance variation acted upon by selection pressure to account for the kind of evolution he envisaged – or for that matter any kind of evolution; because of his need to personify nature in the way he occasionally does.

> "...she may be said to scrutinize with a severe eye ...every habit, instinct, shade of constitution. There will here be no caprice, no favouring: the good will be preserved and the bad rigidly destroyed."

Or in comparing artificial and natural selection:

> "Man selects only for his own good. Nature only for the being which she *tends*... Can we wonder then that nature's productions should be far "truer" in character than man's productions, that they should be infinitely better adapted to the most complex conditions of life, and should plainly bear the stamp of far higher workmanship?"

This is clearly not the language to describe the automatic process of selection Darwin ostensibly proposes where chance vagaries withstand equally chance vicissitudes to fortuitously forge themselves into something (or other). Having subconsciously or consciously grasped the inherent inadequacy of his theory he is forced to invest nature and natural selection with a purpose, a personality, and a power or designing capacity beyond the facts as he states them in order to accomplish what he wants it to.

This is the subtext for *The Origin of Species*, unarticulated for the most part, but unavoidable throughout.

We have drawn particular attention to the extraordinary phenomenon of adaptation which plants and animals so ubiquitously exhibit in relating to almost every environment present on the planet in such a way as to survive. The adaptation of the fish, the bat, the water spider, the bird, the mangrove tree, and the cactus to their respective environments, to take a few creatures almost at random, have been shown to be so exact, so complex, so sophisticated, so perfect that it must be impossible to believe that it has occurred by the accident of chance variation winnowed by selection for survival.

Strictly speaking – which means clearly thinking – even the very concept entailed in the word 'adaptation' cannot be legitimately applied, cannot be used, by upholders of the Darwinian theory. It implies an active, conscious process in which a creature adapts to its situation which is not inherent in the chance process of trial and error which is the limit contained in the theory as it stands. A chance knocking into shape is all it can offer. Yet adherents of the theory habitually use it as though it is in fact an active rather than a passive process, consciously or unconsciously investing it with more than it really contains, or implies. Over and over again we read phrases like "So it adapted to its new environment" or even "it adapted to the new opportunity", and so on...

This is not legitimate. It transgresses the bounds of their theory. It goes beyond it, with a certain sleight of hand.

And the reason is clear. It is a confession of the inadequacy of the theory to do what it is supposed to do. It has to be invested with properties it does not possess: A purpose and a plan. A capacity for design.

Now, though the theory cannot explain it, it is design that we see – unavoidably and ubiquitously. In the intricacy of detail in nature, and in the interwoven harmony of its whole. Workmanship on a colossal scale is resoundingly apparent in the planet we inhabit, and beyond.

Sooner or later we must ask the question – what is it all for? One must turn and face oneself – and stop always being on the run – from oneself and this question. There may be an answer after all. Seek and ye shall find. Don't seek and ye shall not find.

The concept of purpose is not, on the whole, one which is addressed by science in general, or Darwin in particular. His image of nature and living things was horizontal rather than vertical. He had no hierarchy of values. A creature was either simple or complex. Nor did he recognise any 'direction' to the world of life he so carefully studied externally, but not of course beyond that. There was no purpose in his view of the Universe. It was going nowhere, and all that interested him was survival – in other words, existence rather than the 'evolution' he has mistakenly been credited with. Consciousness, that centrally significant attribute of sentient beings, and they all are to a greater or lesser extent, was of no concern to him. It is difficult to avoid describing his as a curiously lacklustre and vacuous approach to life.

This was not true of Darwin's contemporary and much neglected co-founder of the theory of evolution through natural selection, Alfred Russel Wallace. As is not as well-known as it should be, this English naturalist, explorer and later social activist, arrived at the same idea as Darwin at the same time. His paper 'On the Tendency of Varieties to Depart Indefinitely from the Original Type' was in fact presented and read to the Linnean Society of London together with a paper by Darwin on the same theme on the 1st of July 1858 while he was in Malaya. It is well established that Wallace had arrived at the idea independently, having been unaware of Darwin's private research on the subject, and vice-versa.

Unlike Darwin, however, Wallace grew to be dissatisfied with the theory. Though he still believed it could explain some things, there were limits beyond which it could not go. He came to the conclusion that 'something in the unseen Universe of Spirit' had interceded at least three times in the history of the planet.

> 'The first was the creation of life from inorganic matter. The second was the introduction of consciousness in the higher animals. And the third was the generation of the higher mental faculties in mankind.'

He also believed that the raison d'être of the Universe was 'the development of the human spirit.'

Wallace had done two things which were anathema to scientists, and greatly disturbed Darwin who otherwise held him in considerable esteem. He had entered the realm of Purpose; and he was postulating a Universe other than the material. This was heresy of the most extreme order. For which he was strongly criticised by the scientific community, and nearly lost him his pension – as a similar suggestion later lost Fred Hoyle his Nobel Prize. And more recently James Lovelock has been aggressively castigated by the scientific establishment for his concept of Gaia; not only for daring to suggest that the planet Earth is a living being, but also for implying that there is a 'teleological' element in its existence. 'Teleology' is the view that entities and events are due to some purpose or design that is served by them. From the Greek 'telos' meaning 'end'. Most ancient Greek thinkers looked at nature and the Universe in this way.

It is worth noting that Wallace began as an atheist and materialist. But, unlike Darwin's, his conception of evolution evolved. He tells us:

> 'I was a confirmed philosophical sceptic... I was so thorough and confirmed a materialist that I could not at that time find a place in my mind for a conception of spiritual existence.'

He adds:

> 'There was no place in the fabric of my thought into which it could be fitted.'

The phrase 'fabric of my thought' is of course a revealing one. How well it describes what we would now recognise as the shackles of a paradigm. But Wallace was able to break free of it in the course of his development as a man and in his thought, until he came to accept the activity and involvement in nature and the Universe of 'an ever-present, organising and guiding power' as 'an irresistible conclusion.'

A contemporary thinker who would share Wallace's critique of science in general and of Darwinism in particular, is E.F. Schumacher, author of *A Guide for the Perplexed* (1977). His biggest complaint against materialistic scientism is that it rejects the validity of certain questions which for Schumacher are actually the most important questions of all. It would be more accurate for scientists to say that science itself is not valid to pronounce upon questions outside its (limited) scope. In other words, it is not qualified to accept or reject in such areas. It would be truer to say the science rejects its own validity to make statements beyond its legitimate field of operation.

The chief such question that interests Schumacher is that of levels of being. Science lacks the ability to recognise any such concept of nature, man, and the Universe. Darwin, like other scientists, except the very few who are willing to look outside the blinkers of science, do not regard the Universe as a hierarchy of being and consciousness. They have no sense of verticality. Only the horizontal obsesses them – and even that on the lowest of levels.

Schumacher however envisages the Universe as a hierarchy of being. Interestingly, he regards the traditional concept of the Chain of Being as having intrinsic validity as a model of the Universe. He agrees with the view that there are four kingdoms: Mineral, Plant, Animal, and Man, and argues that there are critical and significant differences of kind between each level of being. These are *qualitative* differences. And like Gurdjieff he regards being and consciousness as organically related. However, both according to the original Chain of Being and to Gurdjieff, there are several higher levels of being-consciousness than the human. Though, as has been said, Gurdjieff always insisted that the potentialities of the human level itself has not yet been realised by the vast majority of mankind. Here the Sufis would agree. As they say:

> 'Man has fallen asleep in Life's waiting room. Yet he imagines he is fully alive.'

Or as Gurdjieff puts it: 'At present man lives only in the cellar of his house.'

At this point, if Gurdjieff, the Sufis, and certain other mystical philosophers are right, the question arises: What difference would it make if man did in fact learn to have access to the higher realms of thought and experience – the Sufis call it perception – that they claim exist?

Unless one has attained such a level of intelligence in the truest and deepest sense of the word – and, I regret to say, the present writer has not – it is impossible to say.

Nevertheless, a thorough study of the writings and the sayings of such people allow us to discern a certain feature which appears to appertain to this state which the Sufis call Illumination, and others call enlightenment, and is probably the real meaning of 'wisdom'.

All mystics seem to exhibit what can be described as 'a sense of Unity' – in which they see all things as somehow intricately connected and interrelated in one great Whole. As Happold, that lifelong researcher into the literature of mysticism, concludes:

> 'There is little doubt that a sense of the Oneness of everything in the universe and outside it is at the heart of the most highly developed mystical consciousness. To the mystic is given that unifying vision of the One in the All and the All in the One. Though it may be expressed differently, this is equally true of Hindu and Sufi mystics, of Plotinus, and also of the great contemplatives of Christianity.'

What this means in practice is that certain men and women are able to connect their consciousness in some sense to the Whole, and thus know what is right or wrong in relation to this Whole, what is ultimately the best thing to say or do in a given situation with the benefit of a larger and deeper field of view. It is obvious that mankind needs this faculty, not only to develop, but probably to survive.

We are talking about the need to be aware of the *real* context of anything, the real situation one is in, what could be called the Cosmic Context – an organic relationship with the evolution of the Universe, which includes one's own individual evolution. The touchstone will be spiritual value.

Another way of putting this is the cultivation of a *sense of direction* – how to steer a course through the complex web of life, and to align oneself with its inner flow. And thus ultimately to become the human being one is meant to become.

***

The Ants and the Pen

This allegory, based upon an argument of Rumi's (Mathnavi, IV) was used by the teacher Saad el-Din Jabravi, the founder of the Saadi Sufi. School.

The intention in this version is to admit the usefulness of the scientific ('ant') method of investigation, while insisting that another kind of knowledge ('literacy') not normally associated with man, must be acquired in order to make sense of life.

Jabravi died in Damascus in 1335. His tales are still current, accompanied by the argument that allegory is essential to the human mind to envisage ideas which cannot be captured by any other method.

An ant one day strayed across a piece of paper and saw a pen writing in fine, black strokes.

'How wonderful this is!' said the ant. 'This remarkable thing, with a life of its own, makes squiggles on this beautiful surface, to such an extent and with such energy that it is equal to the efforts of all the ants in the world. And the squiggles which it makes! These resemble ants: not one, but millions, all run together.'

He repeated his ideas to another ant, who was equally interested. He praised the powers of observation and reflection of the first ant.

But another ant said: 'Profiting, it must be admitted, by your efforts, I have observed this strange object. But I have determined that it is not the master of this work. You failed to notice that this pen is attached to certain other objects, which surround it and drive it on its way. These should be considered as the moving factor, and given the credit.' Thus were fingers discovered by the ants.

But another ant, after a long time, climbed over the fingers and realised that they comprised a hand, which he thoroughly explored, after the manner of ants, by scrambling all over it.

He returned to his fellows: 'Ants!' he cried. 'I have news of importance for you. Those smaller objects are a part of a large one. It is this which gives motion to them.'

But then it was discovered that the hand was attached to an arm, and the arm to a body, and that there were two hands, and that there were feet which did no writing.

The investigations continue. Of the mechanics of the writing, the ants have a fair idea. Of the meaning and intention of the writing, and how it is ultimately controlled, they will not find out by their customary method of investigation. Because they are not 'literate'.

From "Tales of the Dervishes"

Idries Shah

Octagon Press

INDEX

RECOMMENDED

LITERATURE

The Sufis

Idries Shah, Octagon Press. The best account of the nature and history of Sufi activity. Source of the story 'The Islanders' in the present work (*Chapter 9*).

The People of the Secret

Ernest Scott, Octagon Press. The learned author develops the concept of an unknown Directorate guiding human evolution.

Stairway to the Stars

Max Gorman, Aeon Press. An introduction to mystical development looking at different esoteric schools.

Tales of the Dervishes

Idries Shah, Octagon Press. A unique collection of stories that teach, transmit, and transform.

Mysticism

F.C. Happold, Penguin Books. A wide-ranging anthology of mystical writings from both the East and the West.